Kitchen Chemistry, Baking Science Why Flour, Sugar, and Butter Do What They Do, Book 2 of 5

Cybersoft Publishing LLC

C. Louis-Charles, PhD

Kitchen Chemistry, Baking Science

This book is intended for general educational purposes only. It is not professional baking, culinary, or food-safety advice and is not a substitute for guidance from a qualified baker, chef, food-safety professional, or healthcare provider. Always handle food safely and follow current food-safety guidance from your local authorities.

First Edition

Published 2026

Contents

Introduction

You pull a cake from the oven, and it has sunk in the middle. You tip bread onto the rack, and the bottom is pale and soft. You bite into a cookie that should have been chewy, only to find something closer to a cracker. None of these moments feels like chemistry failures. They feel like personal failures. You followed the recipe. You measured. You were careful. And still something went wrong inside that sealed box of heat.

This book is about what went wrong and why you can now prevent it.

Baking is the most chemistry-like activity that happens in a home kitchen. That is not a metaphor. It is what baking actually is. When you mix flour, water, and salt, you are building a molecular structure that is locked in before the pan ever reaches the oven. Chemistry is the baking. Baking is chemistry.

I came to this understanding by making mistakes in my own kitchen and then needing to know why. I have a background in chemistry and spend time reading food science research, but the kitchen is where ideas become real. A flat cake is a more vivid teacher than any journal article. This book grew out of those kitchen moments, and it is written for anyone who has had the same ones.

Why Baking Feels Unforgiving

The central frustration of baking is that you cannot taste and adjust as you go. A soup can be fixed at any moment. But once the batter goes into the oven, the experiment is sealed. You find out whether it worked 30 minutes later.

That sealing-in is not a flaw. It is the process. The reason baking feels unforgiving is that the chemistry is set in stone before the heat starts. Every ratio you set during mixing, every temperature your butter was at, is already baked into the result. The oven does not make decisions. You made them earlier.

This sounds discouraging until you flip it around. If the outcome is committed at mixing time, then understanding what happens during mixing is the path to reliable baking. A person who knows why a recipe calls for cold butter handles it at the right temperature without being told. Knowledge is more reliable than careful obedience.

Diagram I.1: The 6-Ingredient Baking Universe. A bold hexagon in flat vector style on a white

That hexagon is the entire universe of baking. Six ingredients. Each one is performing three jobs at the same time. Every chapter in this book peels back one layer of that overlap and shows you what is happening at the molecular level.

Every Ingredient Has Multiple Jobs

The biggest mental shift this book asks you to make is this: no ingredient in baking does only one thing.

Sugar is not in your cookies for taste. It tenderizes the gluten network by competing with the proteins in the flour for water. It breaks down through two separate chemical pathways. It holds moisture after baking, which is why a high-sugar cookie stays soft on the counter for days. In a meringue, sugar is the entire structural material. Eggs are not in a cake just for richness. Egg whites foam and trap air, which lifts the cake. Egg proteins set the crumb structure. Egg yolks carry lecithin, a natural emulsifier that holds fat and water together in a batter that would otherwise break.

Butter is not just flavor. Cold butter in pie dough creates discrete fat pockets that become flaky. Room-temperature butter creamed with sugar creates thousands of tiny air cells. Melted butter produces a dense, chewy structure because no air was trapped. The same ingredient, at three different temperatures, builds three architecturally different results.

This is why baking recipes are more prescriptive than cooking recipes. "Room temperature butter" is a specification about structure. "Ice cold water" in a pie crust is a temperature control that determines whether your fat stays solid or melts into bread. Every instruction that looks arbitrary has a chemical reason. Once you know the reason, the instruction becomes obvious.

Diagram I.2: From Cold Ingredients to Finished Bake. A horizontal four-panel flow diagram with

These four phases (mix, rest, bake, cool) are the architecture of every baked good in this book. Each phase has its own chemistry, and understanding what is happening in each one gives you the ability to intervene when something goes wrong.

How This Book Is Organized

The twelve chapters follow a deliberate sequence. The first seven build your vocabulary of ingredients and processes. The last five apply that vocabulary to specific baked goods and dismantle myths.

Chapters 1 through 7 are ingredient chapters. Flour and gluten come first because gluten is the backbone of nearly everything else. After gluten, the book moves through leavening (the three families of gas production: chemical, biological, and physical), sugar (its functions beyond sweetness), fat (butter at every temperature, plus shortening and oil), eggs (the most versatile ingredient in baking), water and dairy (why liquid composition changes everything), and finally the oven itself.

Diagram I.3: A Baker at the Workbench. A chapter-opener illustration in flat vector style with strong navy outlines

Chapters 8 through 11 are the applied chapters: cookies, cakes, pies and pastry, and breads and laminated doughs. The cookie chapter is a controlled experiment: it uses all six core ingredients and lets you isolate one variable at a time. The cake chapter explains three cake families (butter, foam, oil) and why each handles differently. The pie chapter is a geometry lesson about fat distribution. The bread chapter is about time: how long fermentation takes to develop flavor and how oven steam produces the crust.

Chapter 12 is the capstone. It examines the persistent myths of home baking (always sift the flour, never open the oven door, brown eggs are better, the toothpick test is reliable) and runs each one through the chemistry you have learned. Some myths contain a small core of truth. Most do not. All of them become obvious once you understand the underlying science.

What This Book Promises

By the time you finish, you will be able to look at any baked good that went wrong and name the variable that caused it.

That is a testable promise. A flat cake, a tough crust, a pale loaf, a cookie that spread too much: each of these has a cause, and the cause is always in the chemistry you now understand.

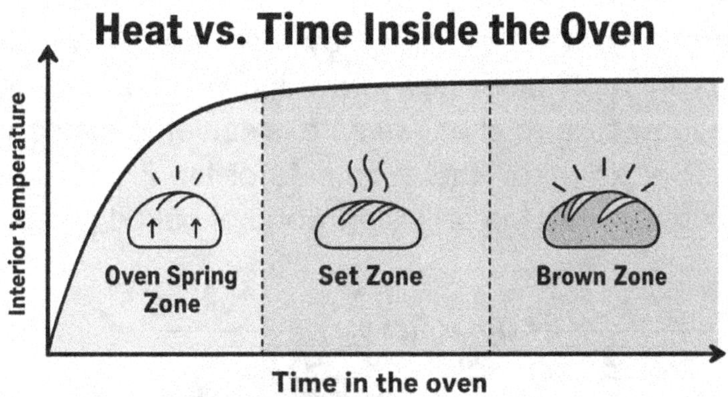

Diagram I.4: Heat vs

The three zones on that graph represent three things that must happen in sequence for any baked good to succeed. Gas expands, and the structure rises. Proteins set, starches gelatinize, and the structure firms. The surface browns and develops flavor. If any zone is cut short, extended too long, or overlaps wrongly, the result fails. Knowing which zone you are in at any moment is the skill this book teaches.

The debugging habit this book builds is direct and kitchen-usable. Did your cookie spread too much because of too much sugar, too much butter, too little flour, or too warm a butter temperature? Each cause feels the same to the person who ate the cookie. But each has a different solution, and you will know which solution to apply.

Where This Book Sits in the Series

This is Book 2 of the five-book Kitchen Chemistry series. Book 1, "Heat, Flavor, Texture," covered the universal science of everyday cooking: how heat moves through food, how browning reactions create flavor, how texture changes across stovetop methods. Book 2 picks up on the oven side of the kitchen and goes deeper into the precision world of baked goods.

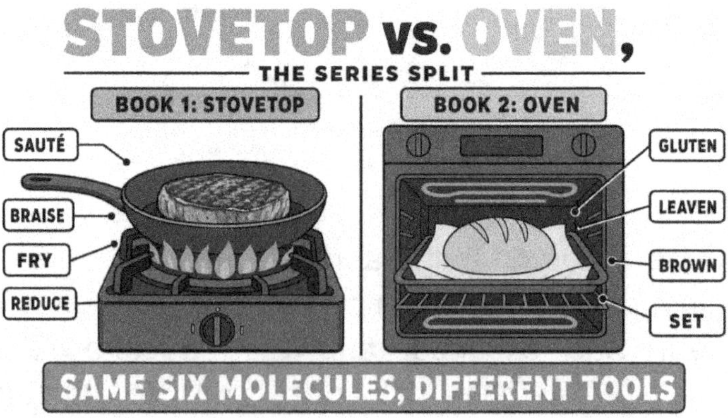

Diagram I.5: Stovetop vs

Books 3, 4, and 5 follow in sequence. Book 3 goes deep into fermentation chemistry: sourdough, wild yeasts, and the bacteria that make sour things sour. Book 4 covers preservation: the science of jam, confit, and pickling. Book 5 is the spice book: why cardamom, cinnamon, and vanilla behave the way they do in baked goods. Each book stands alone. But read in sequence, each one builds on the molecular vocabulary of the last.

The Bold, Modern Bakery

This book is not nostalgic about baking. The image of baking as a gentle, forgiving craft is appealing. It is also not accurate for the person standing in their kitchen at 10 pm with a sunken cake and guests arriving in the morning.

The bakeries that produce the bread people wait in line for (the shatteringly crusted loaves, the cookies with crispy edges and gooey centers, the croissants that separate into 81 layers of buttery flake) are staffed by people who know the chemistry cold. They do not guess. They adjust specific variables to produce specific outcomes. That is the approach this book takes, not sterile or joyless, but clear-eyed and precise.

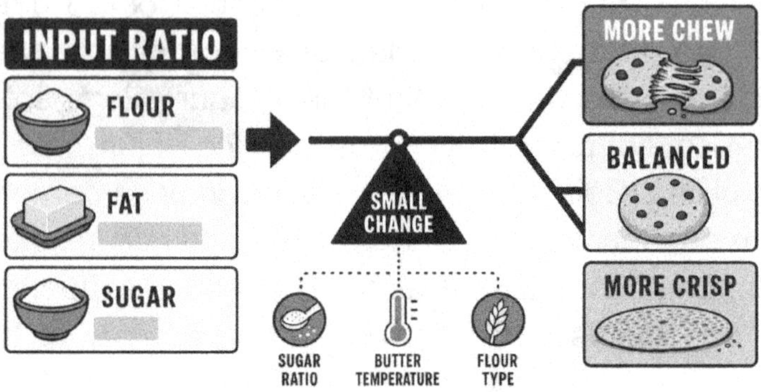

Diagram I.6: The Lever and Outcome Model A flat vector diagram on a white background with strong navy outlines

That lever model is the conceptual spine of every chapter in this book. One variable shifts slightly. The finished texture shifts dramatically. By Chapter 12, you will be reading a recipe and seeing the levers, not just the instructions. You will stop following recipes on faith and start following them with understanding.

What You Can Now Do

You have just read the map. Here is what the journey will give you:

• Identify what each of the six core ingredients is doing in any recipe, beyond what it adds to flavor.

• Recognize which of the three baking phases (rise, set, brown) is responsible when a baked good fails.

• Understand why seemingly arbitrary instructions (cold butter, room-temperature eggs, ice water) are chemical specifications, not fussy preferences.

• Navigate from an ingredient chapter (Chapters 1 through 7) to the applied chapter (Chapters 8 through 11) and connect theory to a specific bake.

• Begin building the debugging habit: name the variable, adjust one thing, and observe the result.

• Approach Books 3, 4, and 5 with a molecular vocabulary that makes each one faster to absorb.

• Walk into any kitchen, look at any recipe, and predict what will happen when that oven door closes.

The chemistry starts in the next chapter. The flour is already waiting.

1 Flour and the Gluten Network

You set a bowl on the counter. You measure out some flour, pour in water, and drag a spoon through it. At first, the mixture looks like library paste: lumpy, wet, unconvincing. You keep stirring. Something shifts. The paste starts to pull away from the bowl in a single mass. You switch from the spoon to your hands. After two or three more minutes of working the dough against the counter, you pinch off a small piece, stretch it between your fingers, and hold it up to the window. Light shines through it like a frosted pane of glass. You just built a gluten network with nothing but flour, water, and your own hands.

That transformation is the central mystery of baking wheat flour. No leavening. No eggs. No fat. Just two dry proteins that were hiding inside the flour, waiting for water to bring them together. The moment water touches those proteins, and you start moving the dough, something begins to form that is neither liquid nor solid, neither stiff nor floppy. It is elastic, extensible, and alive with potential. It is gluten.

Every bread, muffin, pie crust, and pasta noodle you will ever make contains gluten, whether it is intentionally built in or carefully avoided. Understanding how gluten forms, what controls its strength, and how to direct it toward the texture you want is the single most useful thing a home baker can learn. This chapter is your foundation for everything that comes later in this book.

Water plus work builds something from nothing: this is gluten.

Diagram 1.1: The Gluten Window Moment

Why Wheat Is Different

Walk through the baking aisle of any grocery store, and you will see flours made from almonds, rice, coconut, chickpeas, oats, and a dozen other grains and seeds. Each has its fans. But wheat flour sits at the center of Western baking for a reason unrelated to tradition or habit. It has everything to do with two proteins that exist in wheat and almost nowhere else in quite the same form: glutenin and gliadin.

Most plant proteins are roughly globular. They fold into tidy little balls and stay that way. Glutenin and gliadin are different. They are long, coiled, chain-like molecules that want to unfurl and connect to other molecules when conditions are right. When you add water to wheat flour, those proteins absorb it, swell, and begin to reach outward. When you add mechanical energy through kneading or stirring, you force those unfolded chains into contact with each other. They link. They weave. They form something genuinely new: a three-dimensional mesh that is stretchy in one direction and springy in another. That mesh is what we call gluten.

No other common baking flour does this. Almond flour, rice flour, and oat flour contain proteins, fats, and starches, but they cannot build a continuous elastic network. When you mix them with water, you get a paste or a batter, not a dough you can stretch without tearing. This is not a flaw. It is just a different chemistry, one that calls for a different set of tools. We will return to gluten-free baking later in this chapter. For now, focus on what makes wheat flour so remarkable: two proteins that are essentially inert in the dry bag but explosive with potential the moment water arrives.

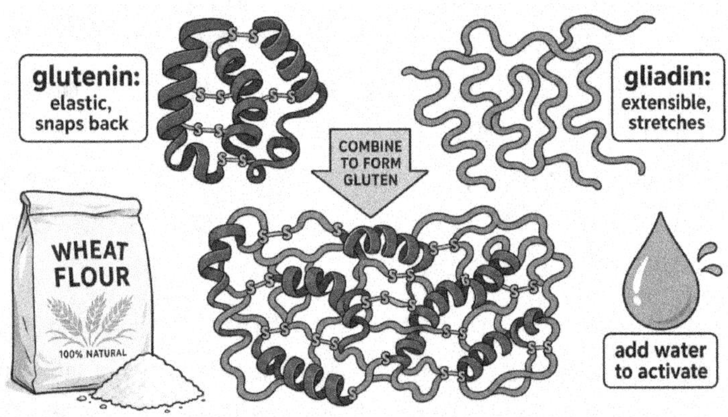

Diagram 1.2: Glutenin and Gliadin Side by Side

The Protein Percentage Story

Not all wheat is the same. Wheat varieties bred for different climates, soils, and end uses have different protein contents, and that percentage is printed right on the flour bag if you know where to look. A high-protein flour contains roughly 13% protein by weight. A moderate-protein all-purpose flour sits around 10% to 11%. A low-protein pastry or cake flour falls around 8%. These numbers sound small, but the difference between 8% and 13% is the difference between a meltingly tender crumb and a jaw-working chew.

Here is why the percentage matters so much. Every additional percent of protein increases the amount of glutenin and gliadin available to build the gluten network. More available protein means more cross-links can form, resulting in a denser, stronger mesh. Bread flour at 13% can build a gluten network strong enough to trap carbon dioxide bubbles through a two-hour rise and a hot oven without collapsing. The mesh is resilient and elastic. Cake flour at 8% builds a much thinner, weaker mesh, one that gives way easily under the pressure of gas bubbles, producing a fine, soft crumb where the cell walls are delicate and thin.

Hard wheat varieties, grown in cold climates with lean soils, tend to be high in protein. Soft wheat varieties, grown in warmer, wetter climates, tend to be lower in protein. When millers label flour "hard red winter wheat" or "soft white wheat," they are summarizing this entire story in three words. The grind matters too: bread flour is typically milled from the whole endosperm of hard wheat, while cake flour is milled more finely and often bleached or otherwise treated to weaken the protein structure further.

All-purpose flour is exactly what its name suggests: a blend calibrated to sit in the middle. It has enough protein to make a decent loaf of bread, enough tenderness to work in most cookie and muffin recipes, and just enough stretch for a basic pie crust. It is not perfect for any one job, but it is good enough for most jobs, which is why most home bakers keep a bag of it on hand and reach for it first.

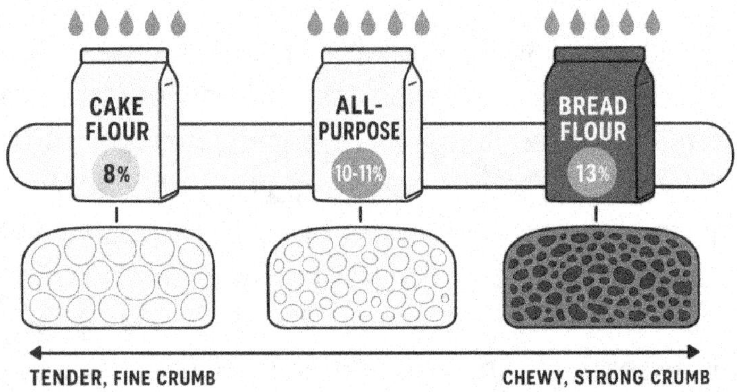

Diagram 1.3: The Protein Percentage Spectrum

Two Proteins, One Network

Understanding glutenin and gliadin separately is useful, but what really matters for baking is how they interact in a hydrated dough. These two proteins are complementary partners. Glutenin provides what chemists call elasticity: the tendency to return to its original shape after being stretched. Gliadin provides extensibility: the tendency to stretch without immediately springing back. A dough needs both qualities. Without glutenin, the dough would stretch and never recover, and your bread would collapse before it finished baking. Without gliadin, the dough would be so stiff and snappy that gas bubbles could not expand at all, and your loaf would be a brick.

When water is added to flour, glutenin and gliadin absorb it through their molecular structure. They swell. Their chains unfurl. Now they are mobile in a way they were not when they were dry. In the dry flour bag, these proteins sit packed together with starch granules, unable to move or connect in any meaningful way. Hydration changes everything. The water molecules work their way between the protein chains, giving them the freedom to shift, rotate, and align.

Mechanical work (kneading, stirring, stretching, and folding) performs the next critical step. It brings protein chains into close enough contact to form new bonds. The most important of these are sulfur-to-sulfur bonds called disulfide bonds. Glutenin is especially rich in sulfur-containing amino acids. When two glutenin chains are forced close together by physical work, those sulfur atoms can snap into a covalent bond that is extremely strong. This is not like two magnets clicking together. It is more like two Lego pieces that lock in so firmly that you need real effort to pull them apart.

The result of all this linking is the three-dimensional mesh we call gluten. In a well-developed dough, this mesh runs in every direction: up, down, sideways, diagonally. It wraps around starch granules and fat droplets. It forms the cell walls that trap carbon dioxide bubbles during fermentation. It gives bread its internal scaffolding. And it holds all of it together from the moment of mixing through the heat of the oven until the moment the bread sets into its final structure.

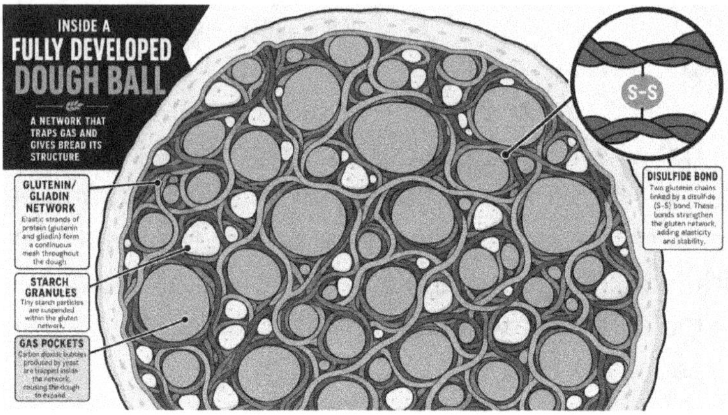

Diagram 1.4: The Gluten Mesh in Cross-Section

Hard Wheat, Soft Wheat, and the Grain Itself

The story of flour protein begins long before the mill. It begins in the field, with the variety of wheat planted and the conditions in which it grows. Two broad categories dominate baking flour production: hard wheat and soft wheat. These are not just marketing terms. They refer to the physical hardness of the wheat kernel, which is directly tied to its protein content and starch structure.

Hard wheat kernels have a vitreous, glassy appearance when cut. The proteins and starches are tightly bound, which is why hard wheat is literally harder to break apart during milling. This tight binding means that hard wheat produces a flour with more concentrated protein and slightly higher starch damage during milling. Starch damage sounds bad, but in bread baking, it is actually useful because damaged starch absorbs more water and gives yeast something extra to feed on during fermentation.

Soft wheat kernels are more opaque and powdery when cut. The protein and starch are less tightly integrated. Milling soft wheat is gentler, and the resulting flour has a finer, silkier texture. This is exactly what cake bakers want: a soft, fine flour that produces a delicate, even crumb. Soft wheat flour also tends to cause less starch damage, meaning it absorbs less water and produces thinner, more pourable batters.

The world's wheat map follows climate logic. Hard wheat varieties thrive in the cold prairies of North America (Kansas, the Dakotas, and Saskatchewan, Canada), where harsh winters stress the grain and force it to develop denser protein reserves. Soft wheat grows more comfortably in the milder climates of the Pacific Northwest, parts of the American South, and much of Europe. When you buy a bag of French T55 flour or Italian 00 flour, you are largely buying soft wheat, milled to a specific fineness. When you buy a bag of King Arthur bread flour, you are buying the output of hard spring or winter wheat.

This geographic logic helps explain why traditional French baguettes and Italian pasta have textures and behaviors quite different from American sourdough. It is not only about technique. The wheat itself is different, and the protein it contains differs in both amount and character.

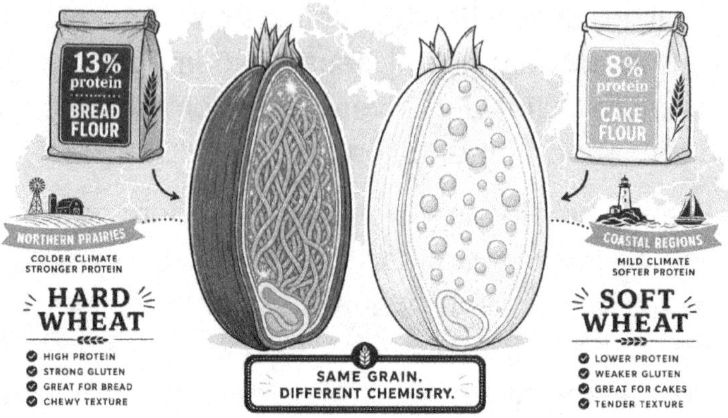

Diagram 1.5: Hard Wheat vs. Soft Wheat Kernel

Hydration: Water Is the Trigger

All the gluten potential in a bag of bread flour means nothing until water arrives. Hydration is the trigger that sets everything in motion, and the amount of water you add has profound consequences for the final texture of your baked good. Bakers often talk about hydration as a percentage of the flour weight. A 65% hydration dough contains 65 grams of water for every 100 grams of flour. A 75% hydration dough is noticeably wetter and stickier. A 50% hydration dough is stiff and almost crumbly before mixing.

When water contacts flour, it is first absorbed by damaged starch granules and then by proteins. The proteins swell as they take up water, and this swelling allows the glutenin and gliadin chains to become mobile. Think of it like this: in dry flour, the protein chains are like a tangle of dry spaghetti crammed into a jar. When you pour water in, the strands soften, swell, and become able to slide past each other. Mechanical work then aligns and links them.

Water also has another job in flour hydration: it activates the naturally occurring enzymes in wheat. Wheat flour contains protease enzymes that gently snip protein chains at certain points. In small amounts, this protease activity is helpful because it loosens the gluten enough to make the dough more extensible. This is part of what happens during autolyze, a technique we will examine in detail shortly. In large amounts or over very long periods, proteases can degrade the gluten network to the point of weakness. This is why extremely long fermentation at warm temperatures can produce a dough that feels like wet cloth instead of elastic dough.

The temperature of the water matters too. Cold water slows both hydration and enzyme activity. This is useful when you want to slow fermentation or give the dough time to rest in the refrigerator. Warm water (around 95F to 100F) speeds hydration and enzyme activity, which can help a stiff, dry dough come together faster. Very hot water (above 140F) can denature proteins, disrupting gluten formation before it even starts. In Book 1's heat chapter, we explored how proteins respond to temperature across a broad range of cooking applications; the same principles apply here to gluten as well.

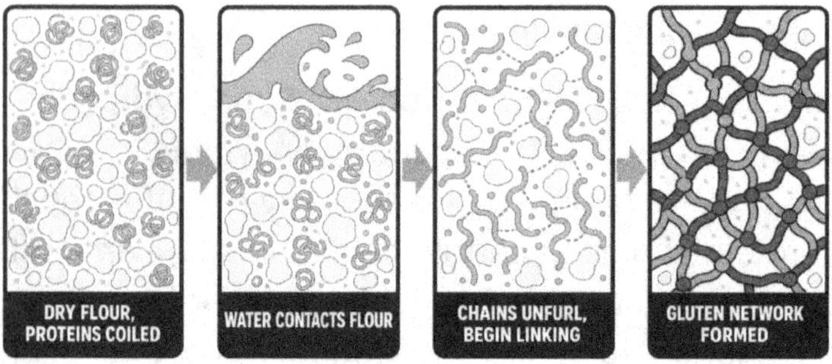

Diagram 1.6: Water Absorption in Slow Motion

Kneading: The Mechanical Builder

If hydration is the trigger, kneading is the builder. Mechanical work transforms a shaggy, uneven dough into a smooth, elastic, cohesive mass. The reason comes down to physics and chemistry working together. When you knead dough, you are doing several things at once: stretching the gluten strands so they align in the direction of force, folding them back on top of each other to create layers of interconnection, and generating enough heat through friction to slightly warm the dough, which speeds the chemical bonding process.

The stretching action is key. When you push a piece of dough away from you with the heel of your hand, you are elongating the gluten strands in one direction. When you fold it back and push again in a slightly different direction, you are layering those strands. Repeat this a few hundred times, and you get a mesh that runs in many directions simultaneously, which is far stronger than strands that all run the same way. Think of a woven fabric versus a bundle of parallel threads. The woven fabric resists tearing in every direction. Parallel threads tear easily if pulled sideways.

You can feel gluten development happening in real time. In the first minute of kneading, the dough is sticky and rough. It tears easily when you pull at it. By the third or fourth minute, it begins to smooth out. By the sixth or eighth minute (for most bread doughs), it has developed a skin-like surface that does not stick to your hands or the counter. It springs back when you poke it. This springiness is glutenin's elasticity expressing itself. The dough remembers where it was and wants to return there.

The ideal endpoint for most bread dough is the windowpane test. You will practice this at the end of the chapter. The idea is simple: if you can stretch a small piece of dough thin enough to see light through it without the dough tearing, the gluten network is well developed. If it tears before it becomes translucent, knead more. This one physical test tells you more about gluten development than any timer can.

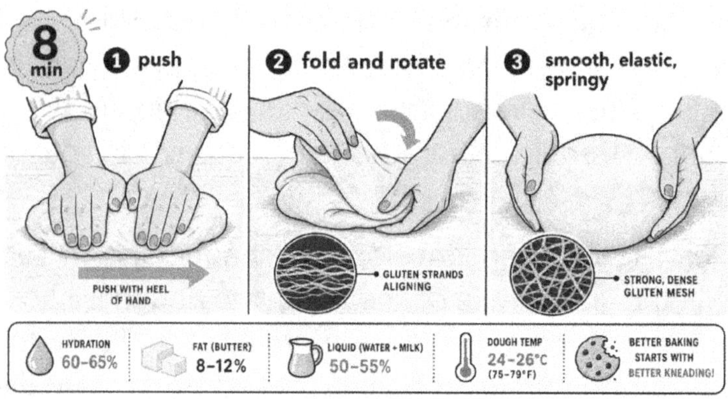

Diagram 1.7: Kneading Mechanics Step by Step

Autolyze: Letting Time Do the Work

There is another way to develop gluten that requires almost no effort: rest. The autolyze technique, widely used in artisan bread baking, involves briefly mixing flour and water, then letting the dough rest for 20 to 60 minutes before adding salt or other ingredients and continuing to work the dough.

During this rest, two things happen. First, hydration continues. Water diffuses slowly into the center of each flour particle, reaching proteins that the initial stir did not fully wet. This more complete hydration means the gluten that forms later will be stronger and more uniform. Second, the naturally occurring proteases in the flour begin to do their snipping work. They cut a small number of peptide bonds within glutenin chains, making the chains slightly shorter and therefore easier to align and link during subsequent kneading. The paradox is beautiful: a little enzymatic breakdown of protein actually leads to better gluten development by making the remaining protein more manageable.

Bakers who use autolyze consistently report that their doughs come together in half the kneading time compared to doughs that receive no rest. The finished crumb is more open and irregular, which many artisan bakers prize. The oven spring (the rapid rise just after the loaf hits oven heat) tends to be stronger as well. This is a technique worth adding to your practice, and it costs nothing but time.

The autolyze is especially valuable with high-hydration doughs. A 75% or 80% hydrated dough is difficult to knead by hand because it is too wet to grip and work against a surface. But if you mix it and let it autolyze for forty minutes, the dough firms up noticeably as the flour fully hydrates. What was a sticky soup becomes a cohesive dough you can actually fold and shape. Time is the only ingredient.

Diagram 1.8: Autolyze Before and After

Stretch-and-Fold: The Gentle Developer

Between the brute force of traditional kneading and the pure patience of autolyze lies a third method that has become a staple of home bread baking over the past two decades: stretch-and-fold. Instead of kneading dough all at once on a counter, you work it in short intervals during the bulk fermentation period. Every thirty to forty-five minutes, you reach into the bowl, grab one side of the dough, stretch it up as high as it will go without tearing, and fold it down over the opposite side. You repeat this for all four sides of the dough, then let it rest again.

The mechanism is similar to kneading but gentler. Each fold aligns gluten strands in a new direction and encourages cross-linking. Because the dough is resting and warming slightly between folds, the gluten relaxes enough to be stretched further with each successive round. By the time you have done three or four sets of stretch-and-folds over the first two hours of fermentation, you have built a strong gluten network without ever working the dough on a counter.

This method is particularly forgiving for high-hydration doughs that would be nearly impossible to knead conventionally. It is also gentler on the developing yeast and bacterial cultures in a sourdough. The simultaneous development of gluten and fermentation activity produces the complex crumb structures and flavors that make artisan sourdough so distinctive.

The stretch-and-fold method also gives you a direct window into gluten development. In the first round, the dough is slack and tears if you pull too hard. By the third or fourth round, it has become noticeably stronger. It resists tearing. It holds its shape better at the end of each fold. You can literally feel the network forming beneath your hands, which is one of the most satisfying parts of baking bread.

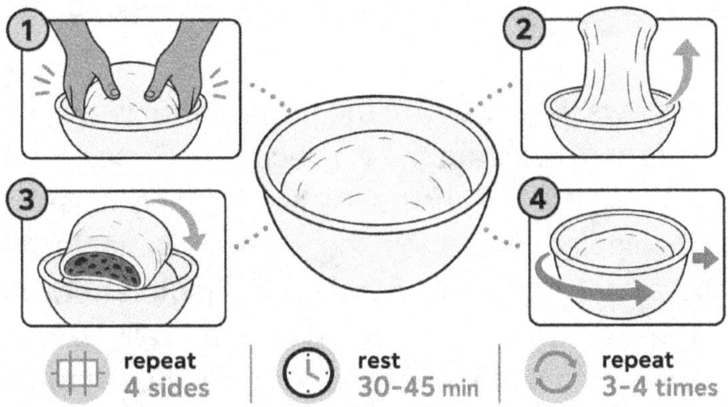

Diagram 1.9: The Stretch-and-Fold Motion

Salt: The Tightener

If hydration and mechanical work are the builders of gluten, salt is the quality-control manager. Adding salt to a dough does something that surprises many home bakers the first time they learn it: it tightens the gluten network, making the dough firmer, less sticky, and more elastic. You can feel this difference directly if you ever mix a no-salt dough and a salted dough side by side.

The mechanism involves ions. Salt (sodium chloride) dissolves in water and releases sodium and chloride ions. These ions interact with the charged groups on protein chains. In unsalted dough, some of those charges repel neighboring protein chains, keeping them from bonding as tightly. The ions from dissolved salt partially neutralize these charges, reducing the repulsion and allowing the protein chains to pack more closely together. The result is a denser, stronger gluten mesh.

Salt also slows the activity of the protease enzymes we discussed in the autolyze section. Recall that proteases snip protein chains. A little snipping is useful. A lot of snipping degrades the gluten. By inhibiting protease activity, salt helps protect the gluten network from breaking down too quickly, especially during prolonged fermentation. This is why yeasted doughs with salt rise more slowly and more predictably than doughs without it: the salt manages fermentation and gluten integrity simultaneously.

The practical upshot is this: if you forget the salt in a bread dough, the finished loaf will be slack, sticky, and pale, and the flavor will be flat. If you add too much salt, you will slow fermentation significantly and produce a very tight crumb. The right amount is about 2% of the flour weight in most bread recipes, a level that has been refined by bakers over centuries and turns out to be almost exactly where the science says it should be.

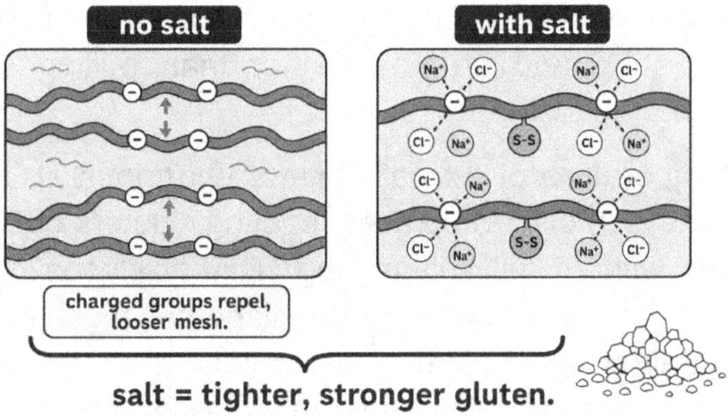

Diagram 1.10: How Salt Tightens Gluten

Fat: The Shortener

Now consider the opposite effect. While salt tightens gluten, fat shortens it. This is where the word "shortening" comes from: a fat that shortens the gluten strands, producing a tender, crumbly texture rather than a chewy, stretchy one. Understanding this one mechanism explains why pastry crust is flaky and tender, why shortbread snaps, and why brioche is so soft despite containing enough flour to build a very strong dough.

Fat works by physically coating the proteins in flour before they can link together. When you rub cold butter into flour for a pie crust, you surround clusters of proteins in the flour with a water-repellent fat layer. Water cannot reach those coated proteins, so they cannot fully hydrate, and they cannot form disulfide bonds with their neighbors. The gluten that forms at all is short, meaning the individual strands are brief and disconnected rather than forming a long, interlocking mesh. When you bite into the baked crust, those short strands snap and crumble rather than pulling and stretching.

The temperature of the fat matters enormously. Cold butter stays in discrete pieces, coating clusters of flour proteins without fully dispersing. This produces flaky layers. Softened or melted butter disperses much more evenly throughout the flour, coating a greater proportion of the protein surface. A dough made with fully incorporated fat (like brioche or a soft sugar cookie dough) has very short, evenly distributed gluten strands, producing extreme tenderness but no flakiness.

Oil behaves differently from solid fat for this reason. Oil is already fully liquid at room temperature, so it coats the proteins in flour evenly from the moment it is added. Olive oil in focaccia, vegetable oil in carrot cake, and coconut oil in quick bread all produce extremely tender crumbs with no flakiness. They are each reducing the gluten as completely as possible. When you choose between butter and oil in a recipe, you are choosing between partial shortening (some flakiness, some chew) and total shortening (maximum tenderness, no flake).

We will go much deeper into fat behavior in Chapter 4, and the specific science of pie crust fat choices gets its own extended treatment in Chapter 10. For now, the key insight is that fat and gluten are rivals, and whoever wins that rivalry determines the texture of your finished bake.

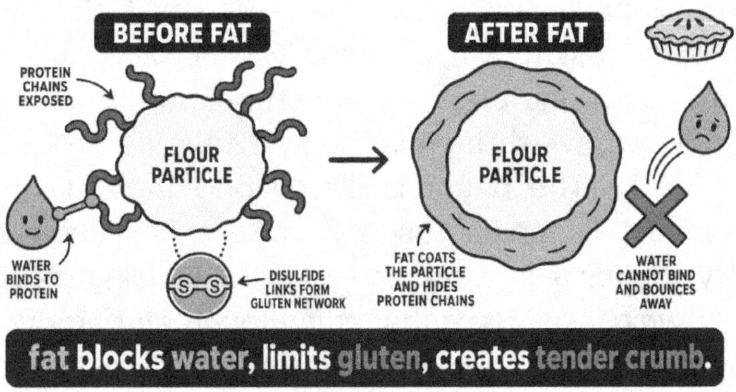

Diagram 1.11: Fat Coating Flour Proteins

The Strong-vs-Tender Spectrum

Every baked good you will ever make sits somewhere on a spectrum from very strong gluten to very tender gluten. Knowing where your target sits tells you everything about how to mix, what flour to choose, and when to stop working the dough. The two poles of this spectrum are the bagel (maximum gluten) and the muffin (minimum gluten). Everything else lives somewhere in between.

A bagel wants extreme gluten development. The recipe uses high-protein bread flour, kneads the dough aggressively for many minutes, adds no fat, and shapes it under tension. The resulting crumb is dense, chewy, and very elastic. You need actual jaw effort to eat a properly made bagel. That is the whole point. The gluten network is so complete and so strong that even the hot water bath before baking (which partially sets the starch) and the high-heat oven cannot fully relax it.

A muffin is the opposite in almost every way. Cake flour or all-purpose flour with a relatively low protein level is used. The fat and eggs are added before the flour is incorporated, which means the fat coats the proteins immediately. The batter is stirred only until the dry ingredients disappear, usually fewer than fifteen or twenty strokes. Any more and you start building visible gluten, which will make the muffin tough and produce a peaked, tunnel-riddled crumb instead of a domed, even one. Quick bread baking is sometimes called "the muffin method," and its central instruction is to stop stirring before you think you need to.

Between bagels and muffins, you have a rich middle landscape. Pizza dough leans more toward the bagel end, but still needs enough extensibility to stretch thin. Baguette dough sits in the upper-middle range, wanting a strong network that can trap lots of gas but still expand to produce the open, irregular crumb prized in artisan bread. Cake layers sit at the tender end, but not as tender as muffins, because the eggs add some structure. Croissants, which we will explore in Chapter 11's laminated dough section, are fascinating because they have a strong gluten network AND a lot of fat, which means the gluten provides structure in one direction. At the same time, the butter layers create flakiness in the other.

Reading a recipe for what it wants from gluten is a skill, and it is one you will develop quickly once you understand the underlying principles. Ask yourself: is this recipe trying to build gluten, limit gluten, or balance gluten? The answer changes everything about how you mix, what temperature you work at, and what flour you reach for.

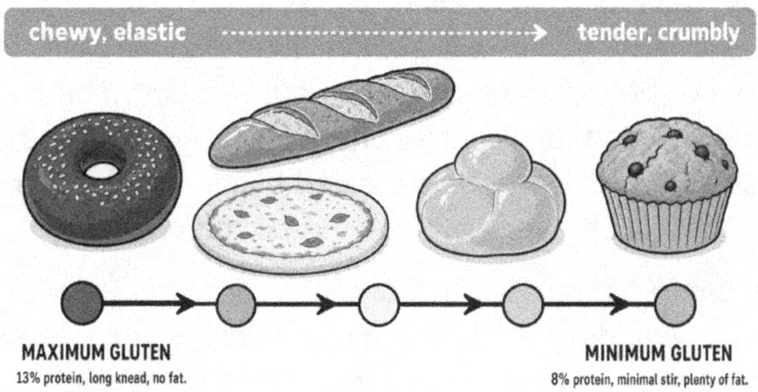

Diagram 1.12: The Strong-to-Tender Bake Spectrum

Why Over-Mixing Ruins a Muffin

Most home bakers have heard the warning: do not overmix your muffin batter. But very few know exactly why. Now that you understand gluten, the explanation is simple and satisfying.

When you add flour to a wet batter and stir, water comes into contact with the proteins, and gluten begins to form immediately. For the first ten or fifteen strokes, you are hydrating the flour evenly and eliminating large dry pockets. The gluten chains are just beginning to link. At this point, the batter is at its most delicate. Stop here, and the gluten will be minimal, the crumb tender, and the dome even and rounded.

Keep stirring past this point, and you are adding energy to a gluten network that is already forming. Each additional stroke encourages more chain alignment, more disulfide bonding, and a stronger mesh. By stroke thirty or forty, you have built genuine gluten in what should have been a fragile, tender batter. The texture of the baked muffin will reflect this: tougher, with a distinctly chewy quality that feels out of place. The top, instead of doming gently, often develops a peaked shape with interior tunnels caused by steam forcing its way through an overly stiff, gluten-bound crumb.

You can easily recover from a slightly undermixed batter by folding it a few more times. You cannot recover from an overmixed batter. Once those disulfide bonds have formed, no amount of rest or gentle handling will break them. This is why professional pastry kitchens sometimes prepare their dry and wet ingredients separately right up to service time, combining them only at the last moment. Timing is everything when gluten control is the goal.

This asymmetry also explains a classic baking question: why do bakery muffins often taste better than homemade ones? Professional bakers use large, high-speed mixers and still stop almost instantly after the flour goes in. Home bakers tend to keep stirring until the batter "looks right," which usually means stirring thirty strokes too many. Restraint at the mixing bowl is the difference.

Diagram 1.13: Undermixed vs. Overmixed Muffin Crumb

Gluten-Free Flours: A Different Architecture

Gluten is such a central part of wheat-based baking that bakers working with gluten-free ingredients often feel like they are trying to build a house without nails. The proteins in almond flour, rice flour, oat flour, and coconut flour do not form a continuous elastic network. They cannot trap gas bubbles the way gluten does. They do not provide the springy, chewy structure that makes bread feel like bread. Working without gluten is not impossible, but it requires a fundamentally different approach to structure-building.

The most common solution is hydrocolloids, which are gummy, water-absorbing molecules that can mimic some of gluten's functions. Xanthan gum and guar gum are the most widely used. Both are polysaccharides (long chains of sugar molecules) that dissolve in water and create a thick, gel-like matrix. This matrix can trap gas bubbles to some extent and give gluten-free batters and doughs a kind of pseudo-elasticity. The result is not identical to wheat gluten, but it is close enough to produce baked goods with an acceptable structure.

Starches also play a larger role in gluten-free baking than in wheat baking. Potato starch, tapioca starch, and arrowroot absorb water and gelatinize during baking, forming a continuous solid matrix as the bake cools. This gelatinized starch network provides the rigidity that holds a gluten-free loaf together after it comes out of the oven. Without it, the loaf would collapse as it cools because there would be nothing to support the crumb structure.

The trade-offs are real and worth being honest about. Gluten-free bread tends to have a tighter, more uniform crumb than wheat bread because the gum-and-starch matrix cannot trap gas as efficiently or expand as freely as a true gluten mesh. The shelf life is often shorter because the starch network can retrograde (re-crystallize) more readily, leading to staling within a day or two. The flavor profile is different because the complex proteins in wheat flour contribute to Maillard browning and flavor development in ways that rice or almond flour cannot fully replicate. None of this means gluten-free baking is inferior. It means it is a different chemistry, with its own set of rules, learning curve, and rewards.

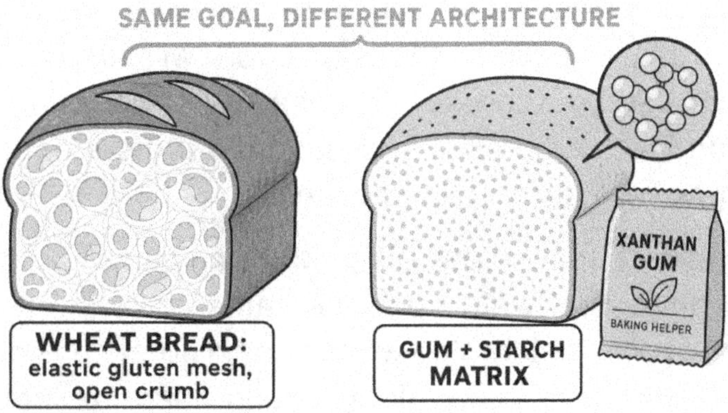

Diagram 1.14: Gluten vs. Gluten-Free Structure

Try This Tonight: The Windowpane Test

This is one of the most satisfying experiments in baking science, and it requires nothing more than two types of flour, some water, and your own hands. It turns an abstract concept (gluten strength) into something you can see and feel in about fifteen minutes.

Start with bread flour. Take a generous handful and add water a little at a time, working it in with your fingers, until you have a firm, non-sticky ball of dough. The dough should feel about as stiff as an earlobe. Knead it on a clean counter for eight minutes. Use the heel of your hand, push away, fold, quarter-turn, repeat. Set a timer if you are inclined. By the end of eight minutes, the dough should be smooth, slightly warm from the friction, and springy when you poke it.

Pinch off a piece about the size of a golf ball. Hold it between the thumb and first two fingers of both hands. Gently, steadily stretch it in all directions, rotating it slightly as you go, as if you were making a tiny pizza disc. A fully developed gluten network will stretch thin enough that you can see the shadow of your fingers through the membrane. The dough is translucent but not torn. That is your windowpane.

Now make a second ball of dough using cake flour, the same amount of water, and the same kneading time. Try the same stretch. You will feel the difference almost immediately: the cake flour dough tears before it becomes translucent. The protein content is not high enough to form a mesh thin enough to stretch. You are not doing anything wrong. You are observing the chemistry of protein percentage in real time.

As a third variation, if you are feeling curious, try a bread-flour dough with two tablespoons of cold butter worked in before kneading. Compare its windowpane to the plain bread flour version. It will stretch, but the window will be cloudier, softer, and less uniform. The fat has done its work as a shortener, even in a high-protein dough. You have just demonstrated, with your own hands, why pie crust is tender, and bagels are chewy.

What you can now do

• Choose the right flour for any recipe by reading its protein percentage: 8% for tender cakes and pastries, 10% to 11% for versatile everyday baking, and 13% for chewy, structured breads and bagels.

• Explain why water is the essential trigger for gluten formation and why the temperature and amount of water you add shape the final dough.

• Use all three gluten development methods (kneading, autolyze, and stretch-and-fold) and choose the one that fits your recipe, your time, and your hydration level.

• Recognize the difference between a properly developed and an overdeveloped dough by feel and by the windowpane test, and stop mixing at the right moment for tender quick breads.

• Understand how salt strengthens the gluten mesh and how fat shortens it, and use that knowledge to predict the texture of a baked good before it even goes in the oven.

• Place any baked good on the strong-to-tender spectrum, from maximum-gluten bagels to minimum-gluten muffins, and read a recipe for what level of gluten development it is asking for.

• Approach gluten-free baking as a different chemistry with its own logic, using gums and starches as structural substitutes and setting realistic expectations for the results.

2 Leavening, How Baked Goods Rise

You scoop the batter into the pan, and it does something surprising. It puffs. It climbs the sides. It rounds into a dome and holds that shape long after it has cooled. Nothing you stirred in looks like a balloon. There is no pump, no fan. Yet the batter has filled itself with gas and set that gas in place. That trick is leavening, and it is one of the most satisfying aspects of baking to understand.

This chapter covers the three families of leaveners: biological (yeast and sourdough), chemical (baking soda and baking powder), and physical (trapped air and steam). Each works through a different mechanism, on a different timetable, and leaves a different fingerprint on your finished bake. Once you know how they work, a recipe stops being a list of instructions and starts being a conversation about which gas is doing the lifting and why.

We explored heat transfer and crust formation in Book 1. You already know that the oven is a hot, dry chamber that drives water out of the surface and sets proteins. Now we will look at what happens before the crust sets, when the dough or batter is still soft, and its gas is still expanding. The race between gas expansion and crust formation is the whole game of leavening.

Diagram 2.1: Batter Rising in a Muffin Cup

What Leavening Actually Is

Before dividing leaveners into categories, it helps to settle on exactly what the word means. Leavening is the introduction of gas into a dough or batter and the trapping of that gas long enough for heat to set the surrounding structure permanently. Every leavened bake has two things happening at once: gas is being produced or released, and the gluten, starch, or protein network around the gas is hardening. If the structure sets before the gas escapes, you get a light, airy crumb. If the gas escapes before the structure sets, you get a flat, dense brick.

The gas itself is almost always carbon dioxide. In a few cases, it is water vapor, which we call steam. In both cases, the principle is the same: a gas is more compressible than a liquid or a solid, but when heated, it expands dramatically. The air we trap when we cream butter or beat eggs is mostly nitrogen and oxygen, but it behaves the same way under heat. Every gas bubble you bake is a tiny balloon. The oven makes it grow; the proteins and starches around it make it permanent.

The structure-setting side depends on what is in your batter. Gluten networks set somewhere between 160F and 180F. Egg proteins coagulate in a similar window. Starch granules gelatinize at around 140°F to 160°F. All of these happen in the same oven at roughly the same time. The shape of a cake is determined in the first fifteen to twenty minutes. After that, the structure is locked. The gas and the protein race each other to the finish line.

Diagram 2.2: Gas Bubble Inside a Battery Network

Baking Soda: Pure Power, No Patience

Baking soda is sodium bicarbonate, nothing else. It is a single white powder that has been sitting in pantries for well over a century. On its own, sodium bicarbonate is stable and does essentially nothing in a dry battery. However, when it comes into contact with an acid in the presence of moisture, it reacts, releasing carbon dioxide gas. That reaction happens fast, which is both its greatest strength and its trickiest characteristic.

The acid in the recipe activates the baking soda. Buttermilk, plain yogurt, sour cream, lemon juice, vinegar, natural cocoa powder, molasses, and brown sugar all carry enough acidity to trigger the reaction. When your recipe calls for baking soda, it is almost always paired with one of these acidic ingredients. The acid and the sodium bicarbonate meet in the batter, react immediately, and release gas. This is why baking-soda batters should go into the oven quickly. The gas is already being produced the moment you stir them together, and if you let the batter sit too long, much of that gas will have escaped before the oven can set the structure around it.

Diagram 2.3: Baking Soda Meets Acid in Batter

Diagram 2.4: Too Much Soda, Side by Side

Baking Powder: Self-Contained and Double-Acting

Baking powder solves the problem that baking soda presents: what do you do when your recipe contains no natural acid? The answer is to build the acid right into the leavener. Baking powder is a pre-made mixture of sodium bicarbonate (the base), one or more dry acids, and a starch buffer. The starch is there to absorb ambient moisture and keep the acid and base separated until they dissolve in the batter. Open the container, and the powder looks uniform and dry. Stir it into a batter, and the starch hydrates, the acid dissolves, and the reaction begins.

Modern baking powder sold in most grocery stores is double-acting. This is one of the most practical inventions in home baking, and it is worth understanding exactly what "double-acting" means.

The powder contains two different acids. The first acid, often cream of tartar or monocalcium phosphate, reacts quickly at room temperature upon contact with moisture. This first burst of gas creates the initial lift you see in batter right after mixing. The second acid, typically sodium aluminum sulfate or sodium aluminum phosphate, does not react much at room temperature. It is heat-activated, meaning it waits for the oven to wake it up. Only when the batter climbs above roughly 140F does the second reaction kick off, providing a second wave of gas right when the batter needs it most, at the start of oven spring.

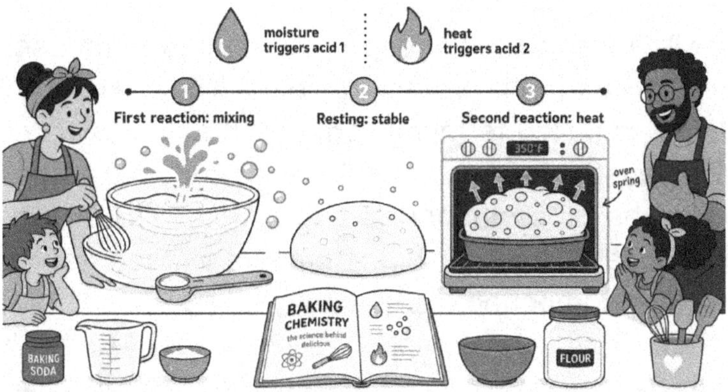

Diagram 2.5: Double-Acting Baking Powder Timeline

Diagram 2.6: Baking Powder Freshness Test

Yeast: The Slow Biological Leavener

Yeast is a single-celled fungus, and it has been used to raise bread for thousands of years. The chemistry it performs is called fermentation, and the short version is this: yeast eats simple sugars and produces carbon dioxide gas plus ethanol as byproducts. The carbon dioxide inflates the dough; the ethanol and the dozens of flavor compounds produced alongside it give yeasted bread its complex, slightly tangy, deeply wheaty flavor that no chemical leavener can replicate.

What makes yeast fundamentally different from chemical leaveners is time. Baking soda reacts in seconds. Baking powder's second wave takes a few minutes in the oven. Yeast works for hours. A standard commercial yeast dough proofs (rises) in one to two hours at room temperature. A slow-fermented dough in the refrigerator might proof over eight, twelve, or even twenty-four hours. This slow timeline is not an inconvenience. It is the source of flavor. The longer yeast ferments, the more organic acids, esters, and other aromatic compounds it produces, and the more interesting the bread tastes.

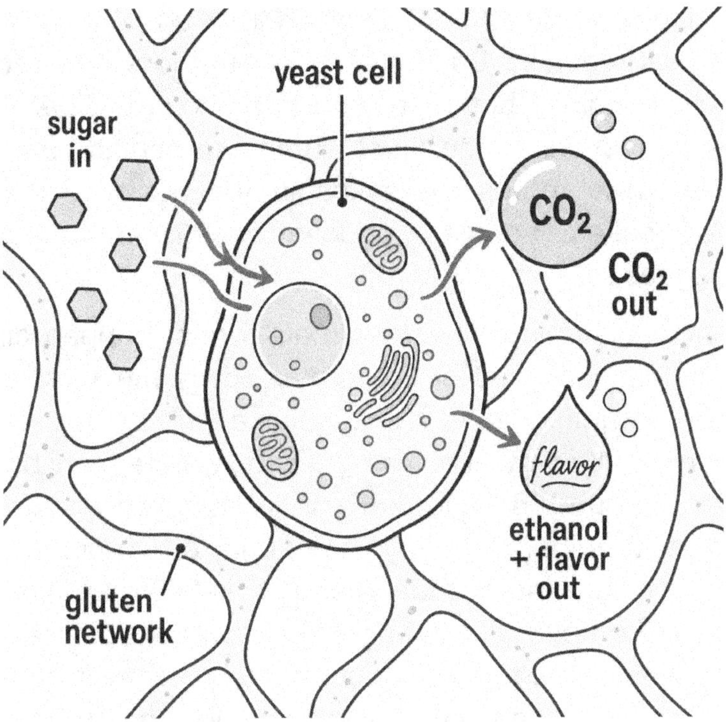

Diagram 2.7: Yeast Fermentation Inside Dough

Diagram 2.8: Yeast Dough Before and After Proofing

Physical Leavening: Air You Build In

Not all gas in baked goods comes from a chemical or biological reaction. Some of it gets put there by hand (or by machine) before baking ever begins. This is physical leavening, the quiet workhorse behind some of the most impressive bakes in the pastry world: chiffon cakes, souffles, popovers, croissants, and meringues.

Physical leavening takes two main forms: trapped air and steam. Trapped air is introduced by mechanical action. Creaming butter and sugar is a form of physical leavening. When a fat at the right temperature (cool but pliable, around 65F to 68F) is beaten with sugar crystals, the sharp edges of the sugar cut millions of tiny air pockets into the fat. These pockets are invisible to the naked eye, but together they form a significant volume of trapped gas. The creamed butter mixture is literally foamy at a microscopic scale. When you bake that cake, the oven heats the air pockets; they expand and contribute lift right alongside whatever baking powder is in the recipe.

Diagram 2.9: Creaming Butter and Sugar

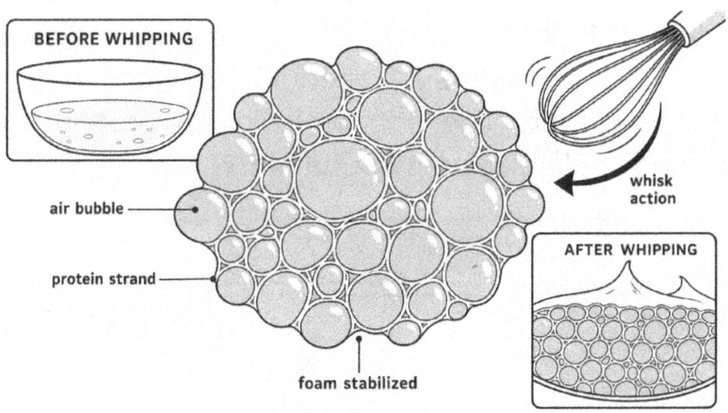

Diagram 2.10: Whipped Egg White Foam Structure

The Power of Steam

Steam is the most dramatic and most underappreciated leavener in baking. Water turns to steam at 212F. When that phase change occurs in dough or batter, it does not just evaporate quietly. It expands with enormous force. One unit of liquid water becomes approximately 1,700 units of steam at atmospheric pressure. That expansion is what makes popovers skyrocket, cream puffs balloon, and croissants spring open in the oven.

Steam leavening works best in structures that have enough strength to capture the steam before it escapes briefly, and enough moisture to produce a large volume of steam, but not so much of a protein network that the structure is rigid before the steam can push it open. Pâte à choux (cream puff paste) is a masterclass in this balance. It is made with a high proportion of water, eggs, and flour, cooked briefly to gelatinize the starch. The starch gelatinization gives the paste just enough structure to hold together in the oven, while the high water content produces a powerful burst of steam that inflates the shell. The egg proteins set around the expanded shell, making it permanent.

Diagram 2.11: Steam Leavening in a Cream Puff

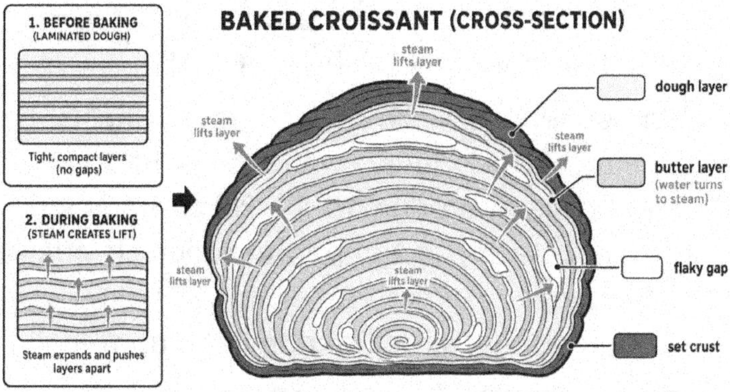

Diagram 2.12: Croissant Layers and Steam Gaps

Why High-Altitude Baking Changes Everything

Bake the same muffin recipe at sea level and at 7,000 feet, and you may get two different results. The muffin at altitude might rise faster, over-expand, and then collapse into a sunken center. The one at sea level holds its dome. Both batters started identically. What changed?

Atmospheric pressure. At sea level, the air above you presses down with enough force to keep gas bubbles relatively compact as they form. A carbon dioxide bubble needs to push against that pressure to expand. At high altitude, the air is thinner, and the pressure is lower, so gas bubbles encounter less resistance. They expand more quickly and to a larger final volume. The batter can rise so fast that it outpaces the rate at which the protein structure is setting. The gluten and eggs have not had enough time to firm up to support the inflated structure, and when the gas finally escapes, or the dough cools slightly, the walls collapse.

Lower atmospheric pressure also lowers the boiling point of water. At sea level, water boils at 212F. At 7,000 feet, it boils at around 199F. This means steam starts forming at a lower temperature, which pushes the steam-leavening effect earlier in the bake, again faster than the protein structure can keep up with.

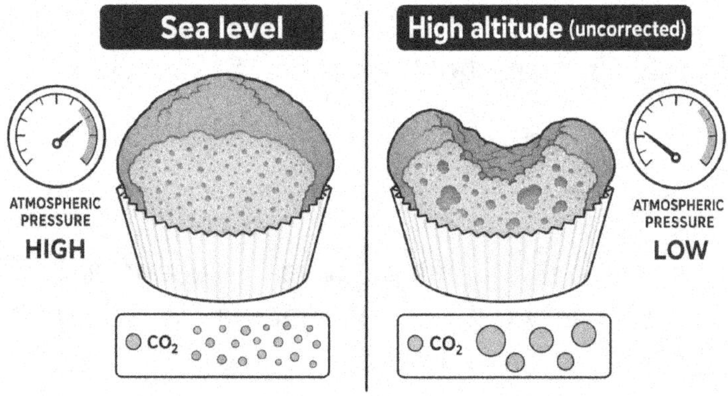

Diagram 2.13: Sea Level vs. High Altitude Muffins

Combining Leaveners: Why Recipes Mix Them

Many recipes call for both baking soda and baking powder. If you are new to baking, this looks redundant. Why use both when either could do the lifting? They serve different purposes in the same recipe, and understanding those purposes lets you decode almost any recipe's leavening line at a glance.

Baking soda is more powerful per gram than baking powder. A small amount does significant work. Baking powder is gentler and more predictable because its acid is built in. When a recipe contains acidic ingredients (buttermilk, yogurt, sour cream, lemon juice, natural cocoa, honey, brown sugar), those acids have to go somewhere. If you use only baking powder, those acids sit unused in the batter and can make the final product slightly sour. Adding a measured amount of baking soda neutralizes some of that acid, balancing the flavor while also contributing extra lift.

At the same time, using only baking soda in an acidic batter is risky because the reaction is fast and aggressive. You may get a big initial rise and then collapse if the structure cannot keep up. Adding some baking powder gives a second, steadier rise in the oven that supports the structure-setting process. The combination gives you the best of both: rapid initial gas from the soda meeting the acid, then a steady supportive second wave from the powder in the oven.

Diagram 2.14: When to Use Soda vs. Powder vs. Both

The Freshness Test and Storage Tips

Two leaveners, two freshness tests. For baking soda: stir a teaspoon into white vinegar. Fresh baking soda fizzes dramatically and quickly. A stale one barely reacts. Baking soda absorbs moisture and odors over time, which reduces its punch. Replace it every six to twelve months.

For baking powder: drop a teaspoon into hot water. Vigorous fizzing means the first acid is still reactive. A powder that fails this test has almost certainly degraded overall. Replace baking powder yearly.

Store both in airtight containers away from heat and moisture. The area near the stove or dishwasher is the worst spot, because steam and heat cycling accelerate degradation. A cool, dry cabinet is ideal.

Yeast has its own freshness test. Dissolve a small amount in warm water (100°F to 110°F) with a pinch of sugar, then let it sit for 10 minutes. Active yeast foams visibly. Dead yeast shows nothing, and the bread will not rise. Instant and active-dry yeast are shelf-stable until opened, then best kept in an airtight container in the refrigerator or freezer. Fresh yeast lasts only one to two weeks refrigerated and should be used close to purchase.

Try This in Your Kitchen

Make two batches of buttermilk pancake batter using your usual recipe. In the first batch, use the amount of baking soda the recipe calls for. In the second batch, use double that amount. Keep everything else identical: the buttermilk, the eggs, the flour, the cooking temperature.

Cook a pancake from each batch in the same pan at the same heat for the same time. Taste them side by side. The first pancake should be tender, golden, and slightly tangy from the buttermilk. The second will be darker (excess alkalinity drives browning) and will carry a noticeable metallic or soapy taste at the back of the palate. That taste is unreacted sodium bicarbonate. No amount of butter or syrup fully covers it.

Cut each pancake open. The correctly-leavened one has a uniform, fine crumb. The over-leavened one has larger, more irregular bubbles and a coarser texture, because the excess gas formed faster than the batter could organize it.

You do not need precise measurements for the second batch. The goal is to experience the off-flavor clearly, not to produce a perfect pancake. When you taste that soapy note, you will remember it. You will recognize it in any over-leavened baked good from that point on.

What You Can Now Do

• Identify which of the three leavening families (biological, chemical, physical) is doing the primary work in any recipe and predict what the crumb texture will look like as a result.

• Use baking soda confidently in recipes that include buttermilk, yogurt, natural cocoa, citrus, molasses, or other acidic ingredients, knowing that the soda needs those acids to react fully and that excess soda will taste soapy.

• Apply the hot-water freshness test to baking powder before any important bake, and replace your supply yearly to avoid flat cakes and quick breads.

• Distinguish single-acting from double-acting baking powder and understand why modern double-acting powders give you a window between mixing and baking without sacrificing rise.

• Cream butter and sugar with intention, knowing that the air you trap in that step is a real and meaningful source of lift in butter cakes and cookies.

• Fold egg white foam gently and purposefully, understanding that each fold either preserves or destroys the air structure that will become the crumb of your angel food or chiffon cake.

• Adjust a recipe for high altitude by reducing leavener, strengthening the batter structure, and raising oven temperature, because you now understand what atmospheric pressure does to gas bubbles relative to protein-setting speed.

3 Sugar, More Than Sweetness

You pull two cookie sheets out of the oven at the same time: same dough, same recipe, same oven rack, same timer. The only thing you changed was the sugar. One batch used all-white granulated sugar. The other got all the brown sugar. And yet the two trays look like they came from entirely different recipes. The white sugar cookies are pale, thin, crisp at the edges, and spread wide. The brown-sugar cookies are puffy, chewy, darker all over, and they hold their rounded shape. You taste one from each tray. Both are sweet. But that is almost the only thing they share.

If you thought sugar was just there to make things taste sweet, this chapter is going to change your mind permanently.

Sugar is one of the most structurally important ingredients in baking. It tenderizes. It browns. It builds and then destroys structure depending on temperature. It holds water, traps air, stabilizes egg foams, and feeds yeast. That last job, feeding yeast and the bacteria that make sourdough sour, is a story Book 3 will go deeper into when it covers the chemistry of fermentation. For now, this chapter focuses on what sugar does in the oven and the mixing bowl. It can be a liquid, a crystal, a syrup, a glass, or a caramel, depending on how hot you take it. In this chapter, you are going to learn what sugar actually does at the molecular level, why different sugars behave so differently, and how to use that knowledge every time you open a bag of flour.

Diagram 3.1: Two Cookie Trays Side by Side

Why Sweetness Is the Smallest Part of Sugar's Job

Most people think of sugar as an ingredient that makes food taste good. That is fair. Sweetness is real, and it matters. But sweetness is the last thing a food scientist puts on the list when describing what sugar does in a batter or a dough. Long before your tongue registers anything, sugar has already been doing structural work: pulling water away from proteins, lowering the temperature at which things set, creating surface area for air bubbles, and setting up the chemical cascade that turns a pale batter into a golden cake.

The key to understanding sugar is to think of it as a molecule that is strongly attracted to water. Sucrose, the molecule in your bag of table sugar, is built from one glucose unit bonded to one fructose unit. Both of those smaller units bristle with hydroxyl groups, which are oxygen-hydrogen pairs that reach out and grab water molecules. This is why sugar dissolves so easily and why it absorbs moisture from the air on a humid day. That water-grabbing behavior is behind almost every functional role sugar plays in baking.

Once you understand that sugar is essentially a water magnet, its five big jobs all make sense. It sweetens, yes. But it also tenderizes by keeping water away from gluten-forming proteins. It browns by reacting with heat in two completely different ways. It stabilizes foams by competing with proteins for water, reinforcing the structure of whipped whites or whipped cream. And it preserves by lowering what chemists call water activity, which is the measure of how much water is available for bacteria and mold to use. Sugar ties up so much water that microbes cannot grow easily, which is why jam, caramel, and royal icing last so long.

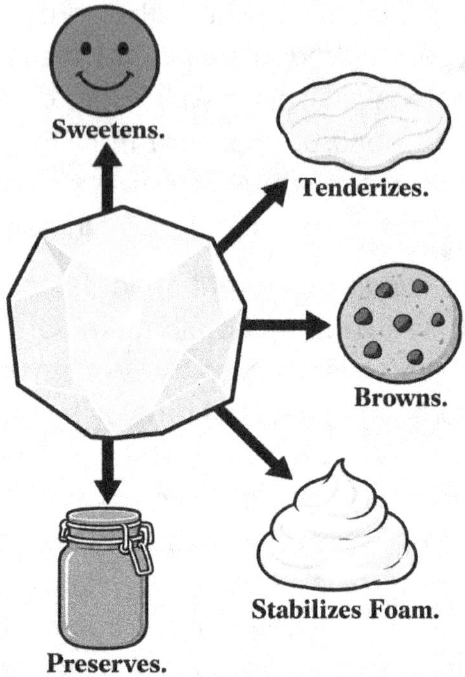

Diagram 3.2: Sugar's Five Jobs

The Crystal You Know: Granulated Sugar

When you reach for the white sugar in your pantry, you are grabbing sucrose that has been refined from sugar cane or sugar beet juice, boiled down, seeded to crystallize, and then dried. Each grain is a tiny crystal, perfectly formed, with flat facets. Those facets matter more than you might expect.

When you cream butter and granulated sugar together, the sharp edges of the crystals are cutting into the fat. Every pass of the mixer blade drives crystal corners into the butter, and each tiny cut traps a tiny air bubble. Those air bubbles are the reason creaming matters. They do not just disappear. They expand in the oven as the gases inside heat up, pushing the batter outward and upward. This is why creaming is actually a leavening step. It is not about flavor. It is about engineering a foam inside your fat before the batter ever goes near heat.

Granulated sugar also dissolves readily in water and liquid, which matters the moment it meets eggs and butter in your batter. As it dissolves, it slightly raises the liquid's boiling point and significantly lowers its freezing point. That freezing-point effect is why ice cream made with more sugar stays softer in your freezer. The dissolved sugar gets between water molecules, making it harder for them to lock into ice crystals. You end up with a creamier, scoopable texture even at freezer temperature rather than a rock-solid block.

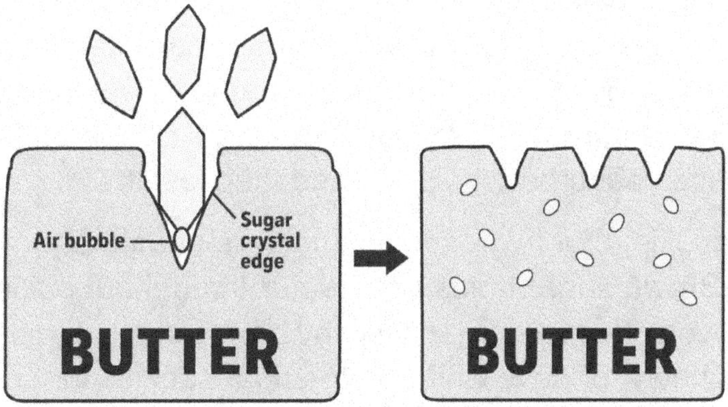

Diagram 3.3: Creaming: Crystal Cuts Butter

Brown Sugar: Molasses Changes Everything

Brown sugar is not a fundamentally different type of sugar. It is white granulated sugar with molasses added back in, or in some cases, never fully removed during refining. Light brown sugar has about five percent molasses by weight. Dark brown sugar has around ten percent. That small percentage makes an outsized difference.

Molasses is the thick, dark syrup that remains after sugar has been crystallized out of cane juice. It is loaded with water, minerals like iron and potassium, organic acids, and complex flavor compounds that survived the refining process. When you add molasses back into white sugar, you get brown sugar that is stickier, clumpier, and more hygroscopic than white sugar. Hygroscopic means it draws water from the surrounding air. If you leave brown sugar uncovered, it will first absorb humidity and then dry hard as a brick when that moisture evaporates. That is why your bag of brown sugar sometimes turns into a solid lump. The molasses absorbed moisture and then lost it.

In a cookie, that hygroscopic quality is a feature, not a bug. Brown sugar holds onto water throughout baking and even afterward, which is why brown-sugar cookies stay chewy for days while white-sugar cookies turn dry and crisp. The molasses also adds acid, which instantly activates baking soda, producing more gas and a slightly puffier cookie. And the extra water in molasses means the dough starts wetter, which slows the cookie's spread in the oven before the structure sets.

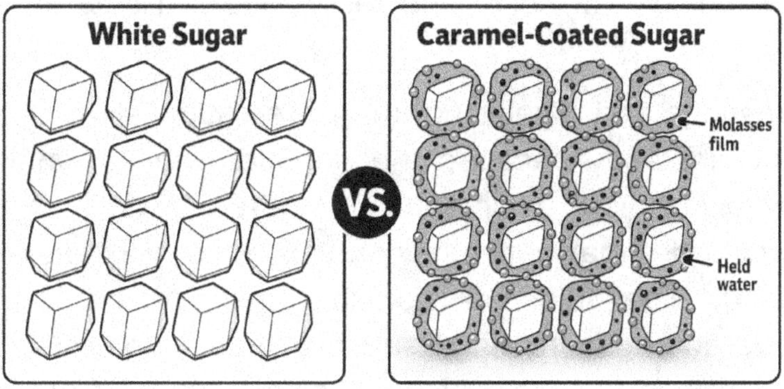

Diagram 3.4: Brown Sugar Molecule Landscape

Powdered Sugar: Why the Cornstarch Is There

Powdered sugar, also called confectioners' sugar or icing sugar, is granulated sugar that has been ground into an extremely fine powder. If you look at the ingredient list, you will almost always see cornstarch listed, usually at around 3% by weight. Many bakers assume the cornstarch is filler or a cost-cutting measure. It is neither.

When sugar is ground very finely, the particles develop an enormous surface area and become highly reactive. They attract moisture from the air with tremendous efficiency and, once they absorb it, dissolve slightly at their surfaces, then recrystallize upon contact with neighboring particles. The result is clumping and, eventually, caking so severe that the powder turns solid. The cornstarch molecules physically separate the sugar particles and absorb surface moisture before the sugar can dissolve and recrystallize. It acts as a buffer.

In frostings and royal icing, powdered sugar dissolves almost instantly because the particles are so small. This gives you a smooth, lump-free texture that granulated sugar can never match without extensive heating. The fine particle size also means that powdered-sugar frostings set to a matte, dry-looking finish rather than the glossy finish you get from a cooked sugar syrup. Royal icing, which is powdered sugar with egg whites and a drop of acid, dries to a hard, ceramic-like shell that is almost entirely sugar glass. When it dries, the sugar has solidified around the air pockets left by the egg white foam.

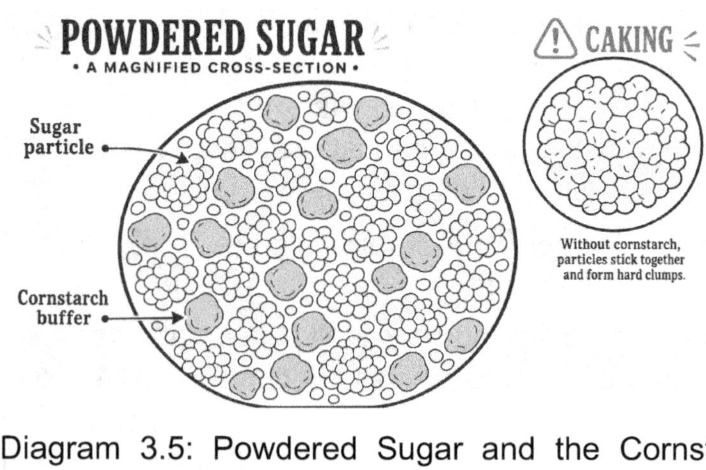

Diagram 3.5: Powdered Sugar and the Cornstarch Buffer

Liquid Sugars: Honey, Molasses, Corn Syrup, Maple

Not all sugars come as dry crystals. Honey, corn syrup, molasses, maple syrup, and invert sugar are all liquids at room temperature, and each behaves differently in baking because of their molecular composition.

Honey is roughly 80% sugar by weight and about 17% water, with the rest being trace compounds such as enzymes, organic acids, and aromatic molecules that give different honeys their distinctive flavors. The sugar in honey is not primarily sucrose. It is mostly fructose and glucose already split apart, because bees add an enzyme called invertase that breaks sucrose into its two component molecules. Fructose is a sweeter-tasting sugar than glucose or sucrose, and it is also more hygroscopic and more reactive in browning reactions. This is why baked goods made with honey brown faster and more deeply than those made with granulated sugar. Fructose reacts more readily with proteins in the batter during the Maillard reaction (which we explored in depth in Book 1's heat chapter), and it also caramelizes at a lower temperature than sucrose.

Corn syrup is different again. Standard corn syrup is mostly glucose, with some longer chains of glucose, such as maltose and other oligosaccharides. The key functional property of glucose syrup in candy-making is that glucose does not crystallize easily the way sucrose does. In a candy or a fudge, if sucrose molecules are the only sugar present, they will tend to find each other and lock into crystals, giving you a grainy texture. Add even a small amount of corn syrup, and the glucose molecules physically get in the way, preventing sucrose from organizing into crystals. The result is a smoother, glossier, more stable candy.

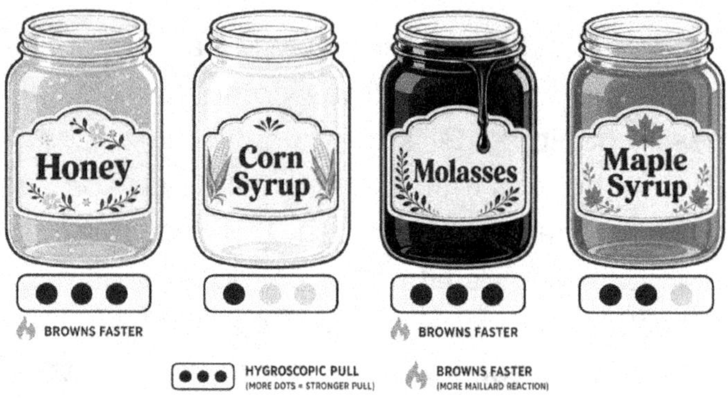

Diagram 3.6: Liquid Sugars Compared

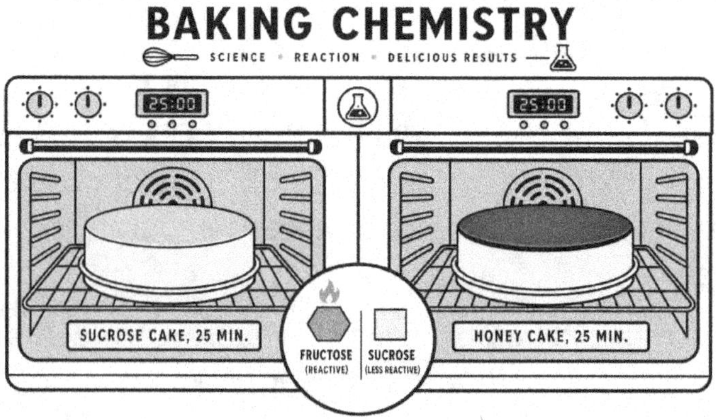

Diagram 3.7: Fructose Browning Faster Than Sucrose

The Candy Stages: Sugar and Temperature

When you dissolve sugar in water and boil the mixture, the water evaporates, and the syrup concentrates. The ratio of sugar to water climbs steadily, and as it does, the physical properties of the syrup change dramatically. This is the foundation of candy-making, but it is also important in baking because caramel, toffee, fudge, nougat, and meringue syrups all rely on understanding what happens to sugar at different temperatures.

At around 234F, a small amount of syrup dropped into cold water forms a soft, pliable ball you can squish between your fingers. This is the soft-ball stage. Fudge and pralines are made here. The syrup crystallizes when agitated, which is why fudge is beaten vigorously after cooling to build a dense mat of fine crystals that give fudge its smooth texture. At 270°F, the syrup reaches soft-crack, used for taffy and nougat. At 300°F, it reaches the hard-crack stage, where it shatters like glass, and the water is almost entirely gone. This is the stage for lollipops and brittle.

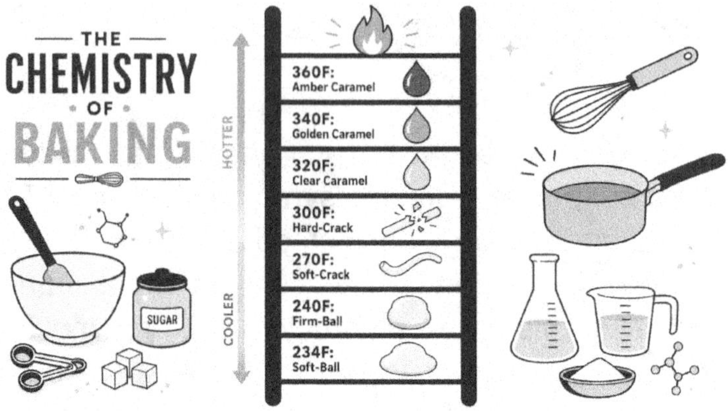

Diagram 3.8: The Candy Temperature Ladder

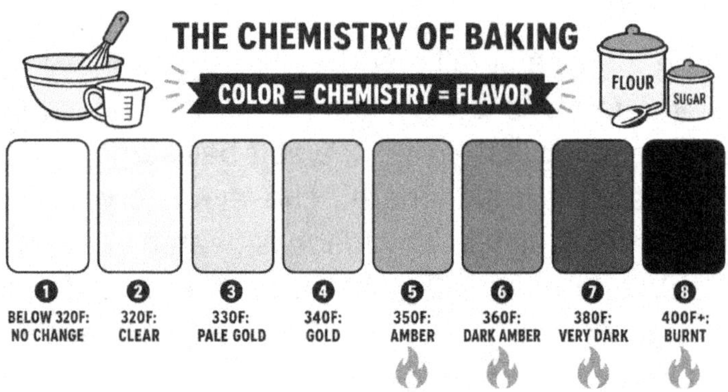

Diagram 3.9: Caramelization Color Strip

Sugar Tenderizes: Competing for Water

One of sugar's most important and least celebrated jobs is tenderizing. When you think of a tender crumb, you probably think of fat. Fat does tenderize, and we will spend a full chapter on that in Chapter 4. But sugar tenderizes by a different mechanism, and the two work together in most cakes.

Gluten, the protein network that gives bread its chew and structure, forms when glutenin and gliadin proteins in flour absorb water and tangle together. The more water the proteins absorb, the stronger and tighter the gluten network becomes. Sugar competes for that water. Because sugar is so intensely attracted to water molecules, it pulls water toward itself and away from the flour proteins. With less water available, the proteins form a weaker, shorter gluten network, and the crumb of the finished cake is more tender.

This is why high-sugar cakes have a noticeably more tender, almost velvety crumb compared to lean bread doughs made from the same flour. Bread dough has very little sugar, so the proteins in the flour absorb as much water as they want and form a tight gluten web. The cake batter has a lot of sugar, so the proteins never get enough water to build that same strength.

This is also why cutting sugar in a cake is more complicated than it sounds. Reduce it by a quarter, and you inadvertently give the flour proteins more water. The gluten network strengthens, and the cake comes out tougher, drier, and less risen. Sugar is structural. Removing it changes the architecture, not just the flavor.

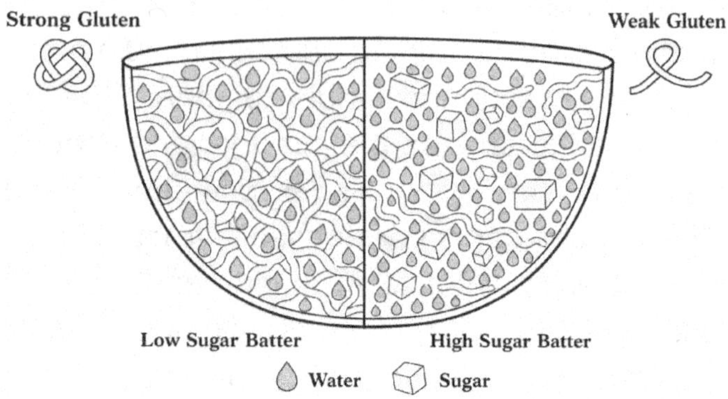

Diagram 3.10: Sugar Competing for Water

Sugar Stabilizes Egg Foams

Meringue is one of the most elegant demonstrations of sugar chemistry. When you whip egg whites, you unfold proteins and trap air bubbles in a foam. Without anything to support it, that foam is fragile: the bubbles gradually coalesce, and the foam collapses back to liquid within minutes. Sugar changes that completely.

When you beat sugar into egg whites, the sugar dissolves in the thin film of water surrounding each air bubble. This concentrated sugar syrup is viscous, meaning it is thick and resistant to flow. That viscosity makes it much harder for the bubble walls to thin and break. The sugar also raises the temperature at which the protein network sets, meaning the meringue can withstand more heat in the oven before it collapses, giving it time to dry out and hold its shape permanently.

French meringue, made by beating raw granulated sugar directly into the whites, is the simplest and least stable type. The sugar takes time to dissolve and may leave fine grains if beaten too quickly. Swiss meringue heats the whites and sugar together over simmering water before whipping, fully dissolving the sugar and denaturing some proteins for a silkier, more stable result. Italian meringue goes furthest: a hot syrup cooked to 240F is poured into the whipping whites, partially cooking them on contact. The result is the most stable, silkiest meringue possible without additives, and it can hold its shape for hours.

Diagram 3.11: Three Meringues Compared

Why Brown Sugar Makes Chewier Cookies

The chewy cookie question deserves a deeper look, because the answer is genuinely chemical. You have already seen that brown sugar contains molasses, which is hygroscopic and holds water. But the full story is richer than that.

The spread of a cookie is a race: how quickly the dough spreads outward as fat melts, versus how quickly the structure sets so the cookie holds its shape. Spread happens before setting. The more the dough spreads before it sets, the flatter and crispier the cookie. Brown sugar slows this race in two ways. First, the molasses moisture makes the dough slightly stickier and more resistant to spreading. Second, the molasses acids react with baking soda immediately, producing some gas early, which puffs the cookie upward rather than outward.

Once in the oven, the hygroscopic molasses does not readily release its water. Even as the cookie heats and the proteins and starches set, the molasses holds onto moisture in the finished crumb. Those retained water molecules are what you feel as chew when you bite in. Chewiness is literally the sensation of water-swollen starch and sugar molecules stretching against your teeth. White sugar dries out completely during baking and lets moisture escape freely, which is why those cookies crisp up. Swap in dark brown sugar or add a spoonful of straight molasses, and you push the chew even further.

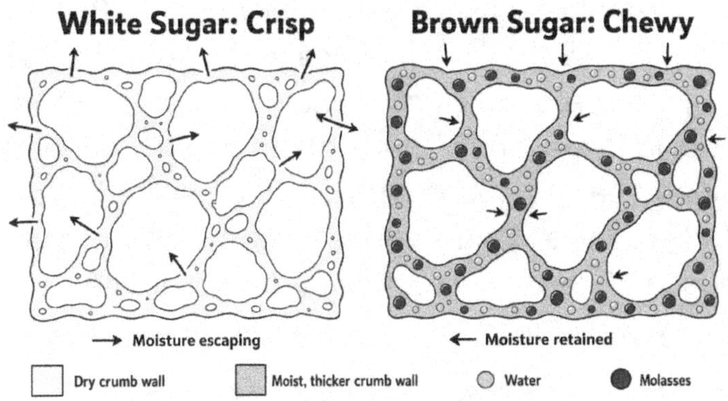

Diagram 3.12: The Chew Mechanism

Honey Browns Faster: Fructose Reactivity

If you have ever substituted honey for sugar in a cookie or a cake, you have probably noticed that it gets very dark very quickly. That is not random. It is fructose chemistry at work.

The Maillard reaction, which we covered in Book 1's chapter on heat, is a reaction between amino acids and reducing sugars. Sucrose in its intact form is not a reducing sugar and does not participate directly. It has to be split first by heat and acid into glucose and fructose before it can react with proteins. Honey, because bees have already split it with invertase, contains free glucose and fructose, ready to react the moment it warms up. More available reducing sugar means faster, deeper Maillard browning.

Fructose also caramelizes at a lower temperature than sucrose. Sucrose begins to caramelize near 320F.

Fructose starts reacting meaningfully near 230F. In an oven set to 350F, a honey cake is both Maillard-reacting and caramelizing at once. A sucrose cake is only Maillard-reacting. That double-browning path is why honey-baked goods turn so aggressively brown. The fix is to lower your oven temperature by about 25 degrees and tent the surface with foil if the top is outpacing the interior.

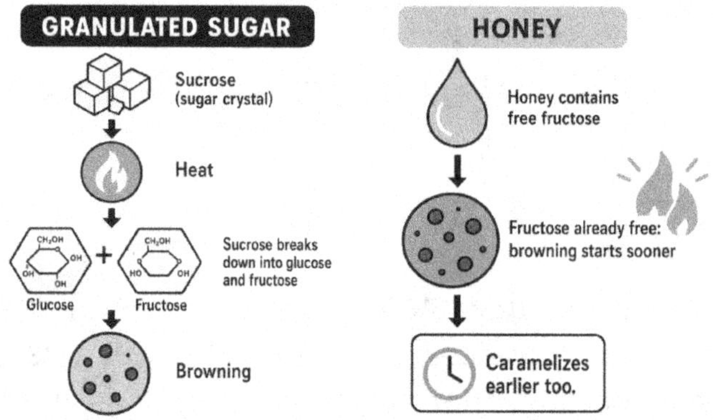

Diagram 3.13: Honey vs. Sugar Browning Paths

Creaming: Air Is the Point

Many recipes begin with the instruction to cream butter and sugar until light and fluffy. Many bakers treat this as a quick step, barely a minute of mixing, before getting to the "real" baking. But creaming is one of the most mechanically significant steps in cake-making, and rushing it changes the result in ways you cannot fix later.

When butter is at the right temperature, around 65F to 68F, it is plastic. That means it can deform under pressure and hold its new shape. Cold butter (straight from the refrigerator, around 40F) is too stiff. The crystals in the fat are too tightly packed to yield to the sugar crystals pressing into them. Melted butter has no structure at all; it is a liquid and cannot hold air bubbles. Room-temperature plastic butter is the sweet spot: it yields enough for sugar crystals to cut in, yet remains solid enough to hold the shape of each cut and the air bubble it creates.

As the mixer runs, each sugar crystal acts like a tiny chisel. It cuts a small groove into the butter surface, and as it passes through, a pocket of air is dragged in behind it. The butter closes around the pocket, trapping the air. Repeat this thousands of times per minute for several minutes, and you end up with butter that has changed color from pale yellow to almost white and has roughly doubled in volume. That increase in volume is entirely air. The butter has not expanded. Air has been forced into it.

Those air bubbles are surrounded by thin, fat walls that help them retain their shape at cool temperatures. When the batter goes into the oven, the air inside it expands with heat, and carbon dioxide from baking powder or baking soda migrates into the existing bubbles, inflating them further. Without those pre-formed bubbles, the leavening gases form new, large, irregular bubbles rather than inflating a uniform array of small ones. The crumb ends up coarser. This is why creaming temperature matters as much as creaming time.

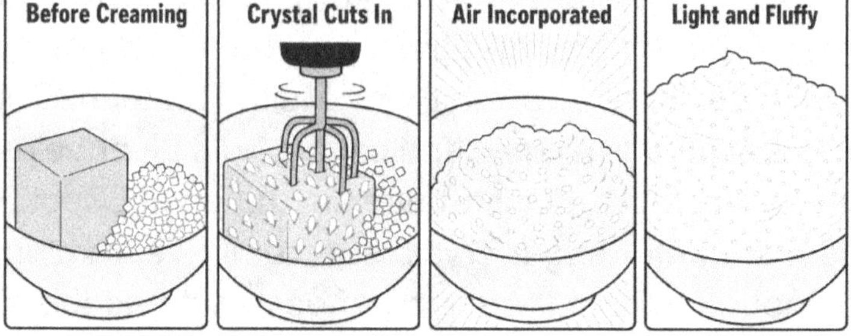

Diagram 3.14: Creaming Process in Four Steps

Try This in Your Kitchen

The best way to feel the difference between sugars is to run the two-cookie experiment yourself. Mix one batch of your usual chocolate-chip cookie dough and divide it exactly in half. Measure the portions by weight to be sure they are equal. Leave one half with all the sugar as granulated white sugar. Replace the white sugar in the other half with the same weight of packed brown sugar. Bake both batches on identical trays on the same oven rack at the same time, or one right after the other on the same rack with no changes to temperature or time.

When they come out, put them side by side on the counter and compare before tasting. Look at the spread: which one is wider? Look at the color: which one is darker? Look at the height: which one is thicker? Press the center gently with one finger: which one gives more and which one springs back stiffer? Then taste them at room temperature and again after they have sat uncovered overnight. The overnight test is particularly revealing. The white sugar cookies will have crisped noticeably. The brown-sugar cookies will remain chewy because the molasses retains moisture.

You have just run a controlled chemistry experiment in your kitchen. You isolated one variable, sugar type, and observed the effect on spread, color, height, texture, and moisture retention over time. Every result you saw corresponds directly to a piece of the chemistry in this chapter.

What you can now do

• Use white granulated sugar when you want crisp cookies, a neutral flavor, and predictable spread; use brown sugar when you want chew, dark color, and a cookie that stays soft the next day.

• Understand that cutting sugar by more than ten to fifteen percent in a cake recipe changes the gluten network and the crumb structure, not just the sweetness, and plan your substitutions accordingly.

• Recognize that honey and other fructose-heavy liquid sugars brown faster than granulated sugar, so lower your oven temperature by about 25 degrees when making a large substitution to avoid over-browning.

• Use corn syrup or invert sugar in candy recipes and frostings when you want to prevent crystallization and maintain a smooth, glossy texture.

• Read the candy temperature stages as a progression of concentrations: soft-ball at 240F for creamy fudge, hard-crack at 300F for brittle, and clear caramel beginning at 320F.

• Control caramel color by watching the color change rather than relying solely on a thermometer: pale gold at 340F, amber at 360F, deep amber at 380F, and smoke means you have gone too far.

• When creaming butter and sugar, start with butter at 65F to 68F, cream for a full three to five minutes, and stop when the mixture looks pale and fluffy; that pale color and volume increase mean the air is in.

4 Fat in Baking: Butter, Shortening, Oil

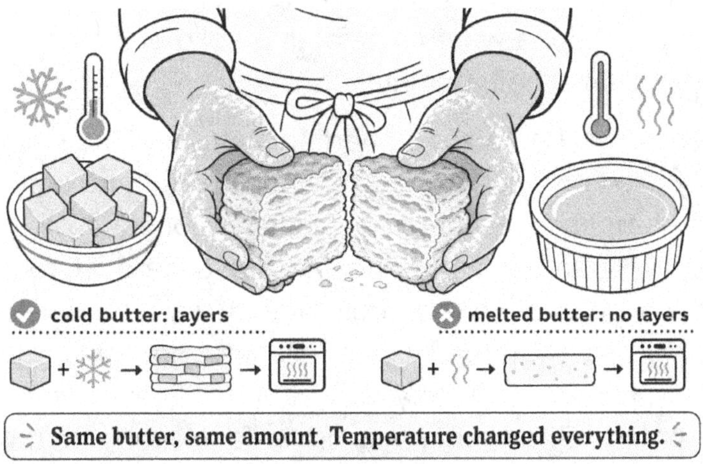

Diagram 4.1: The Biscuit Split, Cold Butter vs. Melted

What Fat Actually Is

Before you can understand what fat does in baking, it helps to understand what fat is at the molecular level. Not in a chemistry-wall way, but in a practical way that will explain every behavior you have ever seen from butter or oil.

All fats, whether butter, shortening, coconut oil, or olive oil, are built from molecules called triglycerides. A triglyceride is a small three-pronged structure, and each prong holds a fatty acid chain. Those chains come in two types. Saturated fatty acids have no double bonds in their carbon chain, so they pack together tightly and hold a firm, solid shape at room temperature. Unsaturated fatty acids have one or more double bonds that introduce kinks in the chain, so they cannot pack closely; they remain fluid at room temperature. The ratio of saturated to unsaturated fatty acids determines whether a fat is a solid or a liquid at the temperature of your kitchen.

Butter sits around 50 to 60 percent saturated fat, which is why it is firm but spreadable when chilled and soft at room temperature. Lard is similar. Coconut oil, at nearly 90 percent saturated fat, stays hard at room temperature and melts sharply when it warms. Olive oil and other plant-based liquid oils are predominantly unsaturated, which is why they pour easily even from a cold bottle. Shortening sits in a fascinating middle ground: it is made from vegetable oils that have been partially or fully hydrogenated, a process that adds hydrogen atoms across double bonds, converting some unsaturated fatty acids into saturated ones and raising the melting point and firming the texture.

That melting point is everything in baking. A fat that melts at 95F behaves differently from a fat that melts at 110F because each gives up its solid structure at different points in the oven's heat. When fat melts early, it becomes liquid and integrates into the dough. When fat melts later, it remains as a separate entity for a little longer and performs different structural work.

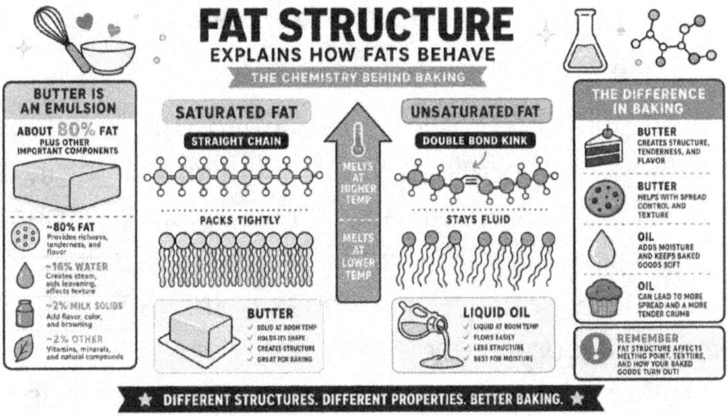

Diagram 4.2: Saturated vs. Unsaturated Fat Chains

Why Temperature Is a Fat's Most Important Setting

Picture three sticks of butter on your counter. You bought them from the same carton. One went straight from the freezer to your cutting board: it is cold, firm, and slices cleanly. One has been sitting on the counter for an hour at about 68F: it is soft and yielding, and it holds an indent when you press it. One melted on the stovetop and cooled to a liquid.

Those three sticks will build three completely different baked goods, even in the same recipe. Not because of flavor. Not because of fat content. But their physical state at the moment of mixing determines what they can and cannot do to the surrounding ingredients.

Cold butter at around 35 to 40F remains solid. When you cut it into flour, the solid pieces coat some flour particles but leave others uncoated. You end up with a mixture that looks like rough, pebbly sand, with visible chunks of fat throughout. Those chunks are crucial. They are the architecture of flakiness.

Room-temperature butter, usually 65 to 70F, is soft but still holds its shape. When you beat it with sugar in an electric mixer, two remarkable things happen at once. First, the sugar crystals cut tiny jagged paths through the butter, and air rushes into those paths. Second, the fat network is supple enough to hold those air pockets open rather than collapse them. The result is a creamed mixture that is noticeably lighter, slightly pale, and full of trapped gas bubbles you cannot see but will feel in every bite of a baked cake.

Melted butter, above its melting point at around 90 to 95F for most butters, is a liquid. When you stir it into flour, it coats every grain of flour evenly and forms a cohesive mass. There are no discrete fat pieces. There are no air pockets. The result is a dough that bakes dense, chewy, and flat. For a fudgy brownie or a thin-and-crispy cookie, that is exactly what you want.

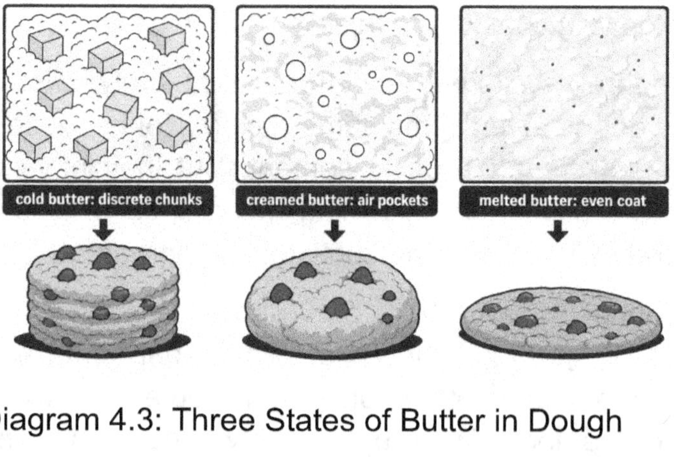

Diagram 4.3: Three States of Butter in Dough

How Fat Shortens Gluten

The word "shortening" is older than the commercial product that now carries the name. It comes from the verb "to shorten," meaning to interrupt, to cut short, to prevent something from extending fully. When early bakers rubbed lard or butter into flour before adding water, they were shortening the gluten, coating the flour proteins with fat so that, when water arrived, the proteins could not fully connect to form long, elastic strands.

Gluten forms when two proteins in wheat flour, glutenin and gliadin, meet water and link up into stretchy networks. Those networks are what make bread chewy and give it the structure to trap gas bubbles during rising. But in a pie crust, shortbread, or biscuit, you do not want those networks. You want tender, crumbly, melt-in-the-mouth texture. Fat is the tool that prevents gluten from doing its job.

When you cut cold butter into flour and work it until the mixture resembles coarse sand, you are coating a large fraction of the flour particles with fat before any water touches them. Those fat-coated particles cannot hydrate properly, so when water arrives later, far fewer gluten bonds form. The result is a fragile, tender network rather than a strong, elastic one.

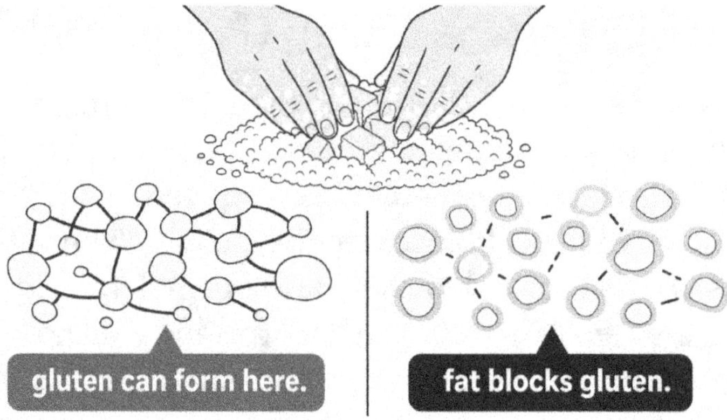

Diagram 4.4: Fat Coating Flour, the Shortening Effect

Cold Butter and the Flake Architecture

A flaky pie crust, a properly made biscuit, a rough puff pastry: these are not mysteries of technique. They are the predictable result of fat remaining solid in distinct pockets during mixing and then doing something dramatic in the oven.

Picture the interior of a raw pie dough as it goes into the oven. You used cold butter, worked quickly, and stopped mixing even though pea-sized chunks were still visible throughout the dough. Those chunks are not accidents; they are a designed feature. Each chunk is a small block of butter: about 80 percent fat and 18 percent water.

When the oven temperature rises, two things happen to each chunk almost simultaneously. The fat fraction melts and spreads into the surrounding flour, creating a local zone that is now water-repellent. And the water fraction (the 18 percent trapped inside the butter structure) turns to steam. Steam is 1,600 times the volume of the liquid water it came from. That explosive expansion has nowhere to go but outward, pushing the dough layers apart.

Diagram 4.5: Inside a Flaky Crust, Steam Lifting Layers

Creamed Butter and Air Pockets

Shift the scene from a cold pastry kitchen to a cake pan waiting on the counter. You are making a layer cake. The recipe says to cream the butter and sugar together for three to five minutes before adding anything else. This step is often rushed. It should not be.

Creaming is not just mixing. It is the inflation of a very precise foam. Butter at room temperature, somewhere between 65 and 70F, has the ideal structure for this. It is soft enough to deform, yet has enough solid crystalline fat to keep the holes that form open. When the jagged edges of sugar crystals cut through soft butter, they create small channels in the fat network. As you continue beating, those channels fill with air.

The evidence is visible and tangible. Start creaming, and the butter looks pale yellow, dense, and a bit greasy. After two minutes, it looks slightly lighter. After four or five minutes of vigorous beating, it is almost white, noticeably fluffy, and holds its shape softly, like very stiff whipped cream. That color change is not cosmetic: it is air. The fat has gone from a dense, packed network to an aerated lattice.

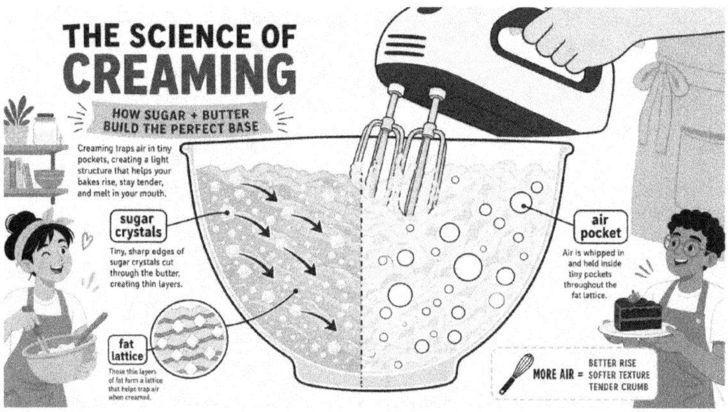

Diagram 4.6: Creaming, Butter, and Sugar Building Air

Melted Butter and the Chewy Spectrum

There is a reason the world's most satisfying chocolate chip cookies are described as thin, chewy, and almost caramelized on the edges. That texture is not an accident. It is melted butter doing exactly what it does best.

When butter is melted before mixing into a dough, it enters the batter as a liquid. Instead of standing apart from the flour, it blends in smoothly and evenly. There are no solid pieces to create discrete fat pockets. No air bubbles form during creaming. The fat coats every particle it touches, creating a cohesive, glossy dough.

What happens in the oven? A cookie made with melted butter spreads more readily because the dough has less internal structure holding it in place. The fat is already in liquid form, so there is nothing to melt out and create the initial resistance you get with creamed butter. The cookie spreads low and wide. With that spreading comes two things: more surface area exposed to direct oven heat, and a thinner dough that dries out and sets faster on the edges while the center stays soft.

Diagram 4.7: Melted Butter Cookie vs. Creamed Butter Cookie

Shortening: Pure Fat, Pure Structure

Walk into any mid-century American cookbook, and you will find shortening listed in nearly every pie crust, cookie, and biscuit recipe. It was the default fat for decades because it delivered reliable, high-volume results with zero fuss. Understanding why it does that will help you decide when it is still the right tool today.

Shortening is 100 percent fat with no water, no milk solids, and no flavor. That composition makes it the most efficient shortening agent available. Because it contains no water, it does not introduce any moisture into the dough. Less moisture means less gluten development, which means a more tender crumb.

Its 100-percent-fat composition also means it holds its solid structure at a higher temperature than butter. While butter starts to soften noticeably around 60F and melts around 90 to 95F, a standard vegetable shortening does not melt until well above 100F. In a hot oven, shortening-based doughs retain their solid-fat structure longer. That delayed melt means more of the dough's shape is set by the time the fat finally liquefies and blends in. The result is cookies that spread dramatically less, and pie crusts that hold their fluted edges precisely.

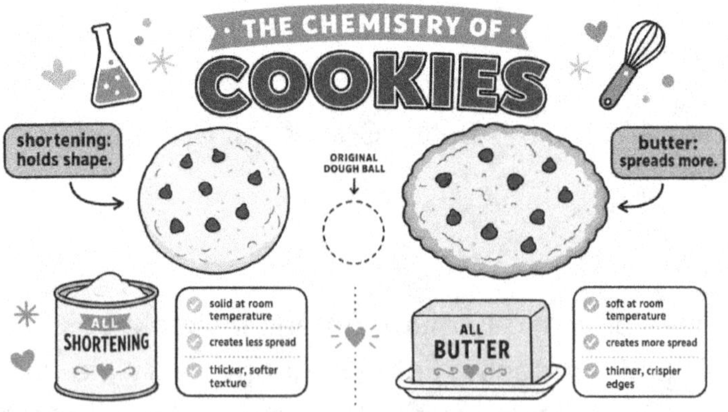

Diagram 4.8: Shortening vs. Butter Spread in Cookies

Oil: Moisture Without Structure

Oil is fat in its simplest form. It is liquid at room temperature, 100 percent fat, and blends instantly and uniformly into any batter. Those properties make it excellent at some things and incapable of others.

The word you hear most often about oil-based cakes is "moist." An oil cake made on Monday is often still moist on Wednesday or Thursday. A butter cake from the same Monday is noticeably drier by Wednesday. Why?

Two reasons. First, oil is a liquid fat at room temperature. When a butter-based crumb sets and cools, some of the fat re-solidifies within the crumb structure, creating a slightly firmer, drier mouthfeel. Oil never solidifies at room temperature, so the crumb stays softer and more tender even after cooling. Second, oil does not hold water in an emulsion the way butter does. Butter's water is bound up inside its emulsion structure, and some of it bakes off during cooking. Oil has no water to lose in the first place. The result is a crumb that retains its perceived moisture better over time.

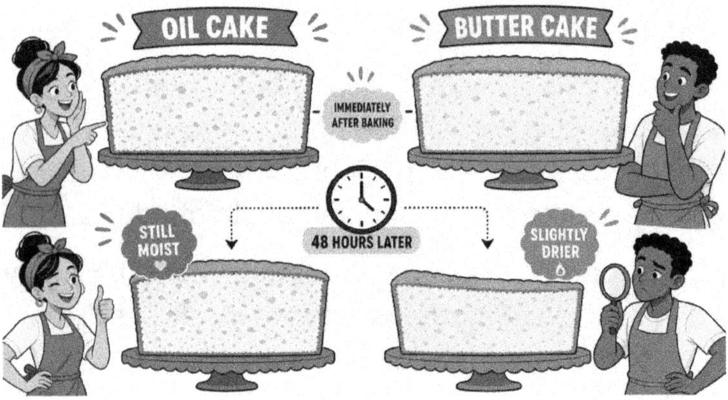

Diagram 4.9: Oil vs. Butter Cake Crumb Over Time

Brown Butter: The Free Maillard Upgrade

One of the easiest ways to make a baked good noticeably more complex in flavor requires no additional ingredients and about eight minutes of active time. You brown the butter.

Browning butter, what French bakers call beurre noisette or literally "hazelnut butter," is a simple process: cook butter in a light-colored pan over medium heat, stirring and swirling, until the milk solids at the bottom of the pan turn golden brown and the whole mixture smells nutty and caramelized. The color transitions in a clear, observable way: melting butter is pale yellow; as the water evaporates, it foams; as the foam subsides and the water is nearly gone, the milk solids begin to brown; the color goes from golden to amber to a rich toasted caramel; the aroma shifts from milky to nutty to almost toffee-like.

What is happening at the molecular level is a Maillard reaction, a reaction in the same family we explored in depth in the heat chapter of Book 1. The milk-solid proteins in butter contain amino acids, and when those amino acids meet milk lactose at high heat, they react to form hundreds of new flavor compounds. Many of those compounds belong to families called pyrazines and furanones, which produce the toasty, nutty, caramel-like notes that make brown butter smell so compelling.

Diagram 4.10: The Brown Butter Progression

Ghee: Brown Butter's Clarified Cousin

Ghee is butter that has been cooked even further than beurre noisette, long enough to remove almost all the water and most of the milk solids, leaving behind a golden, clear, nutty fat that is shelf-stable at room temperature and has a high smoke point. It is traditional in South Asian and Middle Eastern cooking, used both for cooking and for baking.

In baking terms, ghee behaves like a flavorful, 100-percent fat shortening. It has the structural purity of shortening (no water to introduce gluten-promoting moisture) combined with some of the warm, nutty flavor compounds that develop when the milk solids brown before being filtered out. It does not cream well because it has no water-in-fat emulsion to work with, so it is best used in applications that call for melted butter or oil: quick breads, dense cookies, and certain flatbreads.

Because ghee's milk solids have already been removed, it does not produce additional Maillard browning in the oven the way butter does. The browning you taste in a ghee-based baked good happened in the pot before baking. What remains is a fat that carries those flavors but no longer has the proteins and sugars available to create new browning reactions on its own.

Diagram 4.11: Butter to Ghee, What Gets Removed

Saturated, Unsaturated, and Mouthfeel

You can feel the difference between a highly saturated fat and a highly unsaturated one even after the baked good cools. A cookie made with lard or shortening, predominantly saturated fat, feels waxy or clean-cutting on the palate and does not leave a lingering oily film. A cookie made with olive oil or another liquid oil, high in unsaturated fats, feels slightly longer on the palate, with a softer, more lingering richness.

This happens because saturated fats have a sharper, cleaner melting profile. When a saturated fat melts in your mouth, it transitions from solid to liquid over a narrow temperature range, releasing its flavor quickly and cleanly. Many saturated fats melt right around body temperature (about 98F), which means they go from solid to liquid on your tongue very efficiently. That is the sensation food scientists call "clean melt" or a "clean finish."

Unsaturated fats, by contrast, are already liquid at room temperature. They coat the palate more like a fluid than a solid, and because they lack a melting point, they persist as a coating on the tongue and the inner mouth walls. Some palates find this luxurious: the richness of a high-quality olive oil cake lingers pleasantly. Others find it slightly heavy or oily if the fat level is high.

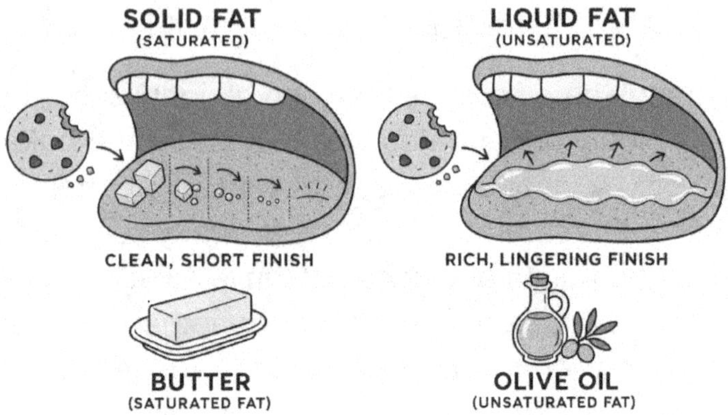

Diagram 4.12: Mouthfeel, Saturated vs. Unsaturated Fat

Why All-Shortening Cookies Do Not Spread

Every baker who has switched from butter to shortening in a drop cookie recipe has experienced this: the cookies come out round, puffy, and tall, sitting exactly where the dough ball landed. They did not spread. The question "Did I do something wrong?" is understandable, but the answer is no. This is exactly what shortening does.

Three factors work together to prevent spreading in an all-shortening cookie. First, shortening has zero water content, so no steam is produced during early baking to push the cookie's edges outward. Second, shortening's higher melting point means the fat stays solid longer during baking. While a butter cookie's fat has already melted and the dough is fluid in the first few minutes, a shortening cookie's fat is still solid, holding the dough's shape. Third, with no water present, the dough is slightly firmer to begin with, less prone to the initial slump that a wet, butter-rich dough might show.

Diagram 4.13: Why Shortening Cookies Stand Tall

Try This in Your Kitchen

The most direct way to see everything in this chapter at once is to run a butter-temperature cookie test. Make a single batch of a simple chocolate chip cookie dough, then divide it into thirds before mixing any fat.

For the first third, cut refrigerator-cold butter into small cubes and work it in just until combined, keeping everything as cold as possible. The dough will be rough and shaggy.

For the second third, use room-temperature softened butter and cream it with the sugar until pale and fluffy before proceeding. The dough will be smooth and soft.

For the third, melt the butter and let it cool to room temperature before mixing. The dough will come together into a cohesive, slightly glossy mass.

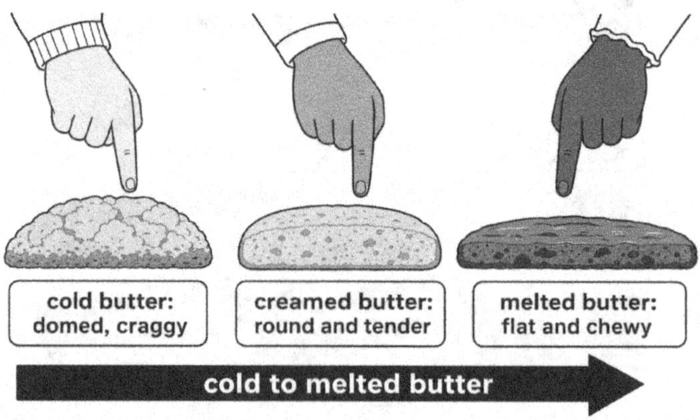

Diagram 4.14: The Butter Temperature Cookie Experiment

Smoke Points: Why They Matter Less Than You Think

The smoke point of a fat is the temperature at which it begins to break down and produce visible smoke. You have probably seen lists of smoke points (butter at roughly 300F, coconut oil at 350F, avocado oil at 520F), usually presented as buying guides or warnings. In baking, these numbers matter less than almost anyone suggests.

Here is why. When fat is in dough or batter, it is not directly exposed to the oven air. It is surrounded by water-containing ingredients, by starch granules, by proteins, all of which insulate it and keep its effective temperature well below the oven setting. A cake baking at 350F will reach an internal temperature of about 200 to 210F when fully baked. A cookie at 375F has an internal temperature around 190 to 200F when it comes out. The fat inside those baked goods never gets anywhere near its smoke point.

Smoke point concerns are particularly relevant when fat is exposed to high heat without an insulating medium, as in pan-frying, deep-frying, or sautéing in an open pan. In those contexts, as we saw in Book 1's heat chapter, choosing a fat with an appropriate smoke point matters. In baking, it rarely does.

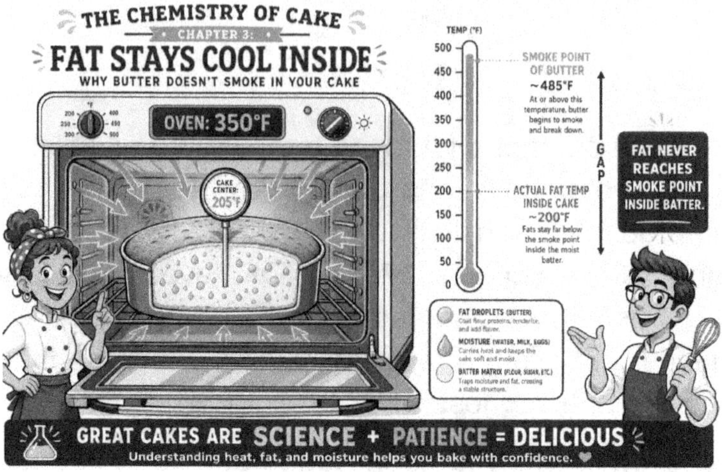

Diagram 4.15: Effective Fat Temperature Inside a Baked Good

What You Can Now Do

• Choose the right fat temperature for the right result: cold butter for flaky layers in pie crust and biscuits, room-temperature butter for creamed cakes and cookies with fine crumb, melted butter for dense and chewy brownies and flat-crispy cookies.

• Understand the shortening effect: fat coats flour proteins and physically prevents gluten strands from linking, which is why more fat and less water in a dough produces tenderness and crumble rather than toughness and chew.

• Make brown butter deliberately: cook butter until the milk solids toast to deep-toasted caramel, then add a small amount of water back if the recipe needs the moisture, and expect a deeper, nuttier, more complex flavor in any cookie, blondie, or cake that uses it.

• Predict how shortening and butter differ in spread and structure: shortening's higher melting point and zero water content mean cookies stay taller and do not spread, while butter's water and lower melting point encourage spreading and browning.

• Know when oil is the better choice: oil-based batters are faster to mix, stay moist longer after baking, and work well with heavy wet ingredients like banana and zucchini, but they cannot cream and will not produce a fine-bubble crumb or butter-brown flavor without help.

• Dismiss smoke-point anxiety in baking: because fat is insulated inside batter or dough, it never reaches its smoke point in a typical oven bake. Select your baking fat based on flavor, water content, and melting behavior instead.

• Use the butter-temperature cookie experiment as your personal calibration tool: bake one batch with cold butter, one with creamed butter, one with melted butter, and let the results teach your hands what no recipe ever quite explains.

5 Eggs, The Ultimate Baking Tool

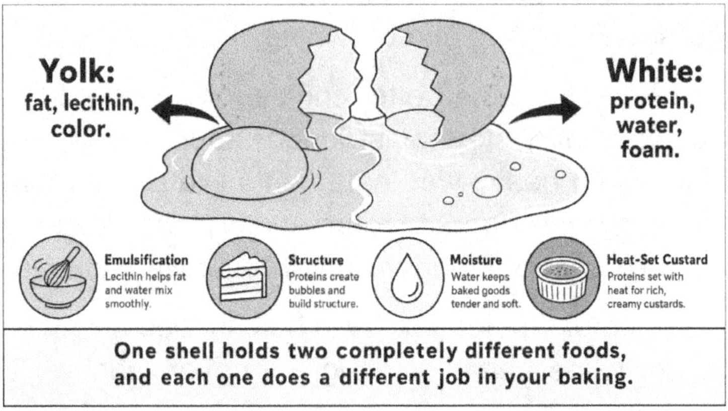

Diagram 5.1: The Two-Food Shell

An Egg Is Two Foods in One Shell

Crack an egg onto a white plate and look at it before you stir it. You are looking at two substances that have almost nothing in common. The white, which food scientists call the albumen, is about 90 percent water and nearly 10 percent protein, with almost no fat. The yolk is a completely different story: roughly one-third fat, a little more than one-sixth protein, a powerful emulsifier called lecithin, and the pigments that give it that characteristic mustard yellow color.

These two components have different textures, flavors, and roles in the oven. The white is thin and slightly viscous when raw, and it turns from transparent to opaque as heat denatures its proteins. The yolk is thick and rich, turning from a flowing liquid to a dense, almost creamy solid as it cooks. When you beat them together into a whole egg, you combine their properties, but you lose some of the precision that comes from working with each one separately.

That is why so many professional recipes treat yolks and whites as distinct ingredients. A genoise whips whole eggs over gentle heat, using both the yolk's richness and the white's protein foam in a single step. A pavlova uses only whites, whipped at room temperature with acid, to build a foam architecture that bakes into a crisp shell with a chewy center. A lemon curd uses only yolks, which contribute fat for creaminess and set at a higher temperature than whites, so the curd does not go rubbery.

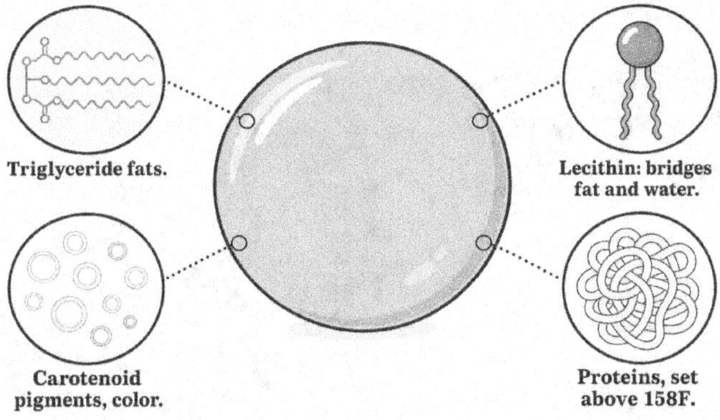

Diagram 5.2: Inside the Yolk - Fat, Lecithin, and Color

The Three Jobs an Egg Does in Baking

Before you can make smart decisions about eggs in a recipe, you need a clear mental map of what eggs are actually being asked to do. There are three distinct jobs, and most recipes ask eggs to do at least two of them at once.

The first job is structure. When egg proteins coagulate in the oven, they form a solid scaffold within the baked good. This scaffold is what makes a cake sliceable rather than pudding-like, what gives a brownie its fudgy bite, and what lets a custard hold a clean spoon-drag line across its surface.

The second job is leavening. Egg whites, when whipped, trap air in a protein foam. That air expands in the oven heat, pushing the batter upward. This is the mechanism behind a genoise, angel food cake, chiffon cake, flourless chocolate cake, and soufflé. In each of these, the baked good rises primarily because of egg foam rather than chemical leaveners.

Diagram 5.3: Three Jobs of an Egg in One Bake

The Egg White Foam and How It Forms

When you start whipping egg whites in a clean bowl with a clean whisk, several things happen in rapid sequence.

First, mechanical disruption folds tiny amounts of air into the liquid. At this stage, the foam is unstable: bubbles are large and irregular, and the mass looks frothy and thin. The white proteins are still coiled in their resting shape.

As you keep whipping, those proteins begin to unfold from the mechanical stress alone, even without heat. Their hydrophobic sections migrate toward the air bubbles and wrap themselves around each bubble, holding it in place. The foam grows denser, the bubbles become smaller and more uniform, and the mass turns from thin froth to thick, glossy, cream-white foam.

Diagram 5.4: Whipping White - Air Bubble Formation

Why Fat Destroys Egg White Foam

Here is the single most important practical rule in egg foam baking: fat is the enemy of egg white foam. Even a trace of fat in your bowl, on your whisk, or in the whites themselves will prevent the proteins from forming a proper foam network.

The reason is molecular. Egg white proteins, when they unfold during whipping, need to wrap tightly around air bubbles and link with neighboring protein chains. Fat molecules interfere with this process. They are slippery; they compete for space at the air-water boundary, and they physically block the protein chains from reaching each other. The result is a weak, unstable foam that deflates quickly and never reaches its full volume.

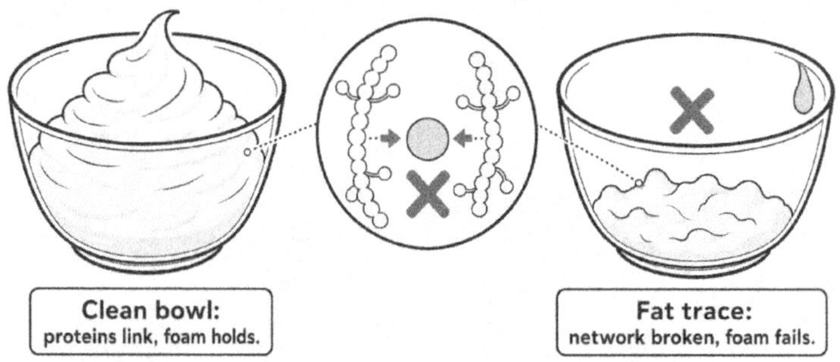

Diagram 5.5: Fat Versus the Foam

Acid, Sugar, and the Stabilized Foam

Once you have a clean bowl and clean whites, you have two additional tools to stabilize the foam: acid and sugar. Both work differently, and both matter.

Acid, most commonly cream of tartar, lowers the pH of the white, which changes the electrical charge on the proteins. At a lower pH, the protein chains are slightly more unfolded even before whipping begins, so they wrap around bubbles more easily and link together more readily. The slightly acidic environment also slows over-coagulation, widening the margin between perfect stiff peaks and the grainy, weeping over-whipped stage.

Room-temperature whites whip faster and to greater volume than cold whites because the proteins have more mobility at warmer temperatures.

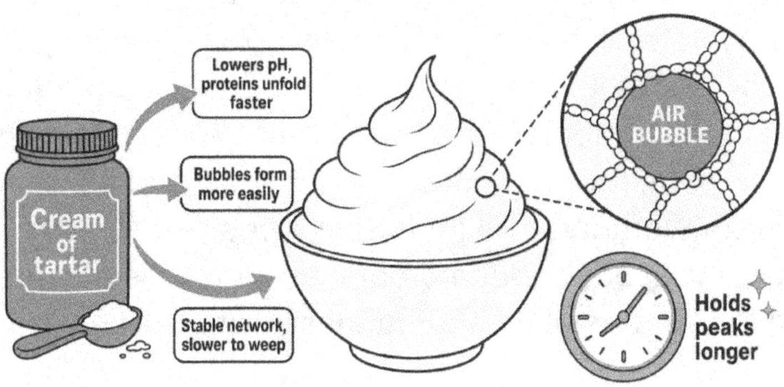

Diagram 5.6: Cream of Tartar and Foam Stability

The Three Meringue Methods

Not all meringues are made the same way. There are three main methods, each producing a foam with different properties, different uses, and different stability. Understanding the chemistry behind each helps you choose the right technique for the job.

The French meringue is the simplest. You whip raw egg whites with granulated or powdered sugar until stiff peaks form. The result is a light, voluminous foam that is ready quickly but is the least stable of the three. It is best used immediately: piped and baked into meringue cookies, spread over a pie, or folded quickly into a batter. French meringue is the most fragile because the protein network has not been heat-set.

The Italian meringue uses hot sugar syrup instead of dry sugar. You cook sugar and water to around 240F (soft-ball stage), then pour the hot syrup in a thin stream into whites already at soft peaks. The hot syrup partially cooks the whites as it hits them, denaturing some of the proteins immediately and creating a firmer, denser foam. Italian meringue is the most stable: it can sit at room temperature for hours, pipes cleanly, and is frequently used in buttercream and in desserts that need to hold shape without further baking.

Diagram 5.7: Three Meringue Methods Side by Side

Yolks as Emulsifiers in Cake Batter

In cake baking, the yolk's most important job is the one lecithin handles: keeping fat and water mixed in a stable emulsion throughout the batter and through oven heat.

A butter cake batter is a fat-in-water emulsion. Butter, when creamed with sugar, creates tiny air pockets and fat droplets that must stay suspended in the liquid portion of the batter. A broken emulsion produces a greasy, separated batter that bakes unevenly. The yolk's lecithin prevents this.

The order of ingredient addition in a butter cake recipe is not arbitrary. When a recipe tells you to add eggs one at a time to a creamed butter-sugar mixture, it is asking the lecithin in each egg to incorporate before the next egg fully destabilizes the mixture. Adding all the eggs at once can overwhelm the lecithin in the batter, causing it to look curdled and broken. Adding them slowly lets each yolk do its emulsification work before the next arrives.

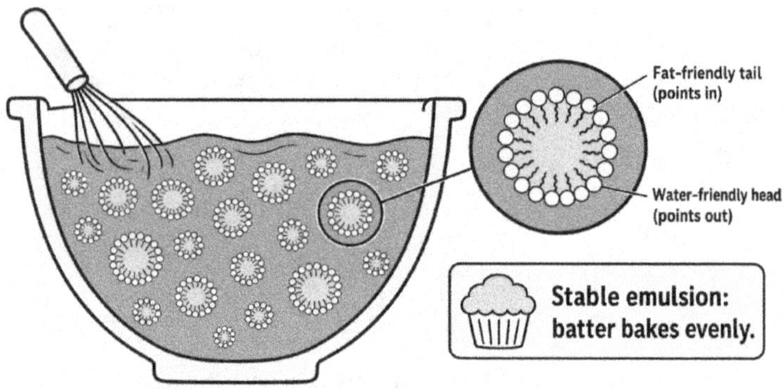

Diagram 5.8: Lecithin at Work in Cake Batter

Custards, Curds, and the Critical Temperature Range

Custards are where egg chemistry is at its most precise and unforgiving. The difference between a silky, spoonable vanilla custard and a bowl of sweet scrambled eggs is a matter of five degrees.

Egg yolk proteins begin to coagulate around 158F. Egg white proteins start lower, around 144F to 149F. When you make a stirred custard like crème anglaise or lemon curd, you are cooking the yolks in a water- and sugar-rich liquid. The sugar and extra liquid slightly raise the coagulation temperature, giving you a larger working window. But it still demands your full attention.

A custard cooked to 175F to 180F is silky and smooth. The yolk proteins have coagulated just enough to thicken the liquid into a flowing, spoon-coating consistency. At 185F, the proteins over-coagulate and begin to squeeze out the water between them. The custard breaks into a curdled, grainy liquid with floating yellow clumps. Sweet scrambled eggs in a saucepan.

Diagram 5.9: The Custard Temperature Ladder

Genoise, Whole-Egg Foam, and the Heat Method

The genoise illustrates a principle that is easy to miss: whole eggs, not just whites, can be whipped into a foam. You do not need to separate eggs to trap air.

The challenge with whole-egg foam is that the yolk's fat disrupts the protein network the same way any other fat does. Whipping whole eggs cold produces weak, low-volume foam. The solution is heat.

When you warm whole eggs with sugar over a double boiler, whisking gently until the mixture reaches around 110F to 115F before transferring to a stand mixer and whipping to full volume, the heat does two things. First, it slightly thins the fat, making it less likely to disrupt the protein network. Second, it partially denatures the proteins, giving them a head start on unfolding so that when mechanical whipping begins, they wrap around air bubbles faster and more effectively than cold eggs would.

Diagram 5.10: Genoise - Warm Egg Foam Technique

Why Room-Temperature Eggs Matter

If you have ever made a butter cake and noticed the batter looking curdled or broken halfway through mixing, cold eggs are the most common culprit.

A butter cake batter emulsion depends on the butter fat being soft and workable. When you add a cold egg to a warm butter-sugar mixture, the cold egg drops the local temperature of the fat. Cold fat seizes up and becomes stiff. Stiff fat cannot efficiently emulsify the water in the egg, causing the emulsion to break. The batter looks curdled, with visible globs of butter floating in liquid.

A broken batter can sometimes be rescued by adding a tablespoon of sifted flour and mixing gently. But a batter that was never broken in the first place bakes better.

Diagram 5.11: Cold Egg Versus Room-Temperature Egg in Batter

Eggs in Cookies and Brownies

Cookies and brownies are where the egg variable becomes the most clearly observable, because the recipe quantities are small enough that swapping one egg component for another produces a dramatically different result in a single bake.

A standard chocolate chip cookie made with one or two whole eggs provides moderate structure, richness, and chew. Make the same cookie with an extra yolk and no extra white: you have added fat, lecithin, and a small amount of extra protein. The extra fat tenderizes the crumb, creates a chewier, moister center, and keeps the cookie soft longer at room temperature because the emulsified fat locks in moisture.

Add an extra white instead of an extra yolk to the same cookie: you have added protein foam capacity and water without adding fat. The cookie spreads more, finishes crispier at the edges, and dries out faster. It is a great cookie for dunking in coffee, but it is not the soft, chewy version you would choose for a gift tin.

Diagram 5.12: Egg Swaps in Cookies and Brownies

Substituting Eggs and What You Give Up

Sometimes you need to replace an egg entirely because of an allergy, availability, or dietary choice. It is worth being honest about what you are giving up and how different substitutes compensate for different jobs.

A commercial egg replacer typically uses a blend of starches and raising agents. It does a reasonable job replacing the binding and structural role of eggs in cookies and quick breads, a poor job replacing foam-leavening, and essentially nothing to replace emulsification. A genoise or angel food cake made with commercial egg replacer will be flat and dense.

Flax eggs (a tablespoon of ground flaxseed mixed with three tablespoons of water, left to gel for five minutes) work well for binding in dense baked goods like muffins, banana bread, and some cookies. The mucilage that forms when the flax soaks up water is viscous and sticky, acting as a binder. It contributes no foam or emulsification and a slight nutty flavor.

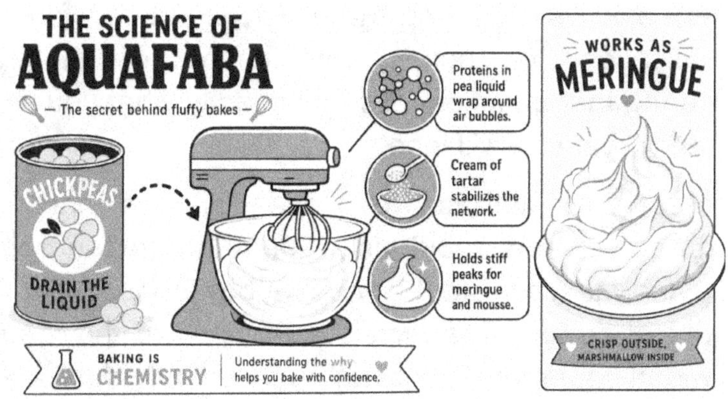

Diagram 5.13: Aquafaba - the Surprising Foam Substitute

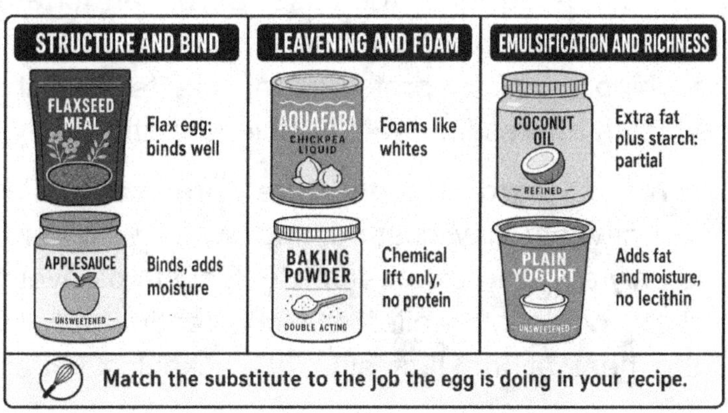

Diagram 5.14: Egg Substitution at a Glance

What You Can Now Do

You have walked through an egg from shell to molecular structure, from foam to custard, from meringue chemistry to substitution logic. Here is what you can do with that knowledge today.

• **Identify the job before you change the formula.** Before swapping whole eggs for extra yolks or extra whites, ask: Is this egg here for structure, foam, emulsification, or some combination of these? Answering that question tells you exactly what you will gain and what you will lose.

• **Protect your whites.** Keep every surface that will touch egg whites scrupulously clean and grease-free. Use glass, stainless steel, or copper bowls. Add a small amount of cream of tartar early to stabilize the foam and widen your margin between perfect stiff peaks and over-whipped disaster.

• **Use heat to unlock whole-egg foam.** When a recipe calls for whole-egg foam (genoise, European tortes), warm the eggs and sugar together over a double boiler before whipping. Warm protein unfolds faster, and the thinned fat interferes less with bubble formation.

• **Control custard temperature precisely.** The window between silky custard and sweet scrambled eggs is only about five degrees. Cook custards over a water bath or double boiler, stir constantly, and pull from the heat at the first sign of thickening. Strain immediately into a cold bowl if you see any hint of curdling.

• **Bring eggs to room temperature before baking.** Cold eggs break butter cake emulsions and whip to lower volumes. Thirty minutes on the counter changes the chemistry of your bake in ways that are clear in the finished crumb.

• **Match your meringue method to your application.** French meringue is fast but fragile: use it immediately. Italian meringue is the most stable: use it for frosting and fillings that need to hold for hours. Swiss meringue is the versatile middle ground for frosting, pavlova topping, and smooth, torchable finishes.

• **Substitute with precision, not volume.** When eggs cannot be used, identify the job first: flax eggs for binding, aquafaba for foam, extra fat plus starch for emulsification. No single substitute covers all three jobs, so matching the substitute to the specific role the egg plays is the only reliable path to a workable result.

6 Water, Milk, and the Liquid Architecture

You pull two nearly identical cakes from the oven. The milk cake is a deeper golden brown, almost caramel at the edges. You cut in, and the crumb is softer, more pillowy. The water cake is fine, pleasant. But the milk cake is something else entirely. You made no other change. Same flour, same sugar, same eggs, same baking powder. All you changed was the liquid.

Now imagine a third cake made with buttermilk. It rises higher than either of the first two. The crumb is almost impossibly tender, slightly tangy. The baking soda, which did almost nothing in the water cake, worked furiously here because it finally had an acid to react with. Three liquids. Three completely different results. Same recipe.

Liquid is not a neutral ingredient in baking. It is structural. Once you understand what water does, what milk adds on top of it, what buttermilk's acid contributes, what cream's fat concentration changes, and what scalding milk was really doing, you stop swapping liquids randomly and start choosing them with intent.

Why is liquid structural

Water is the ingredient that makes gluten possible. Without it, the proteins in flour sit dry and inert. Add water, and two proteins, called glutenin and gliadin, begin to hydrate and link into the elastic, stretchy gluten network we explored in Chapter 1. That network holds gas bubbles, traps steam, gives bread its chew, and gives cake its crumb.

Water also dissolves sugar, salt, and chemical leaveners. Salt does not work until it is in solution. Baking soda does not release carbon dioxide until it meets an acid in solution. Every chemical reaction in a batter or dough happens in the water phase. Without it, nothing reacts.

And water turns to steam. When dough or batter hits a hot oven, the water inside vaporizes. That steam inflates air pockets, pushes up crumb structure, and in pastry, forces flaky layers apart. We covered steam as a leavener in Chapter 2. Here, the point is simpler: the amount of water in your dough controls how much structure-building happens, how much reactivity happens, and how much steam-lift happens.

Diagram 6.1: The Liquid Architecture - Chapter Opener

Hydration percentage: the baker's measuring stick

Professional bakers constantly talk about hydration percentage. It sounds technical. It is actually one of the most useful numbers you can know, because it tells you immediately what kind of dough you are dealing with.

Hydration percentage equals the weight of water divided by the weight of flour, multiplied by 100. If you have 500 grams of flour and 325 grams of water, your hydration is 65%. Flour is always 100% in baker's math. More water relative to flour means a wetter, more open, more extensible dough. Less water means a stiffer, denser, more easily shaped dough.

At 60% hydration, you get a stiff dough. Think bagels. A bagel dough at 60% is firm enough to roll by hand into a rope, shape into a ring, and hold on a board without spreading. The relatively low water content means less steam in the oven and a thicker, chewier crust. That chew is the whole point.

At 65% hydration, you are in sandwich bread territory. The dough is soft and pliable but holds its shape in a loaf pan. The crumb will be soft and even, the kind that slices cleanly and holds a sandwich filling without tearing. At 75% hydration, the dough is noticeably wetter. Think ciabatta. A 75% dough slumps when you try to shape it. Bakers use stretch-and-fold instead of traditional kneading because the dough contains too much water to knead conventionally. That extra water creates the open, airy crumb with large holes.

Push hydration to 80% or above, and you are in focaccia territory. The dough behaves more like a thick batter. You pour it into an oiled pan, dimple it with your fingers, and bake it flat. The high water content produces abundant steam, giving focaccia its airy, spongy crumb.

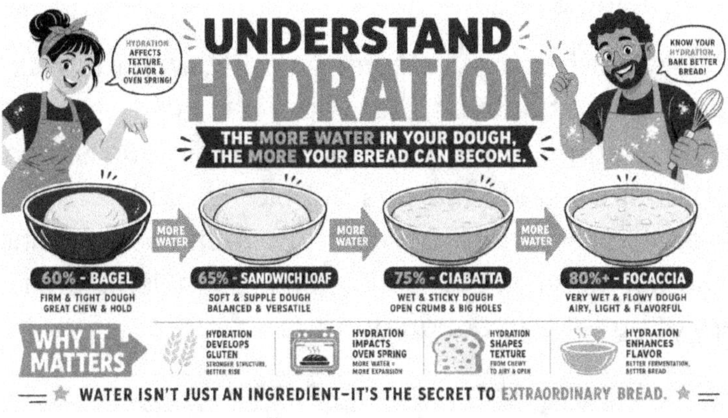

Diagram 6.2: Hydration Spectrum - Stiff to Pourable

Water's basic jobs in a batter or dough

Water has four core jobs in baking. First, it hydrates protein. The gluten-forming proteins in flour need water to swell, link, and form the elastic network that gives baked goods structure. No water, no gluten.

Second, water dissolves. Sugar dissolves in water, evenly distributing sweetness and browning potential throughout the batter. Salt dissolves in water, which allows it to strengthen gluten and season the dough all the way through. Baking soda and baking powder dissolve in water, activating them. Yeast needs water to rehydrate from its dormant dry state and begin producing carbon dioxide.

Third, water hydrates starch. Flour starch granules absorb water and swell in the oven, a process called gelatinization. Gelatinization begins around 140F. The starch thickens and sets, giving baked goods their interior structure and preventing collapse when you pull them from the oven.

Fourth, water becomes steam. At 212F, water turns to vapor. Inside a hot oven, that vapor has nowhere to go except up, through the network of protein and starch. In bread, it expands the gas bubbles already created by yeast or baking powder. In puff pastry and croissants, steam is the primary lift mechanism, forcing layers of butter and dough apart.

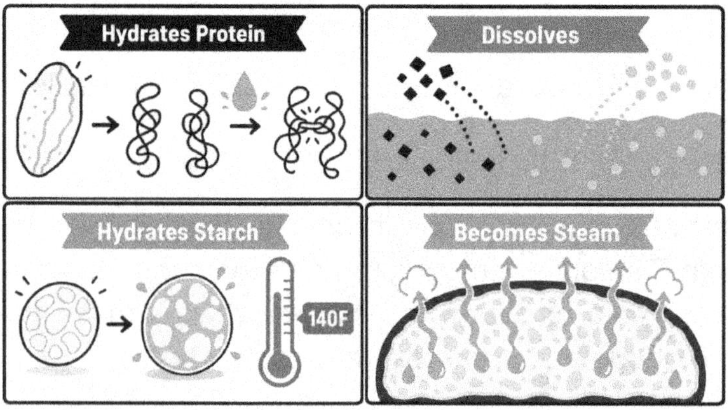

Diagram 6.3: Four Jobs of Water in a Dough

Milk: water with passengers

Think of milk as water with a complex cargo. By weight, whole milk is about 87% water, 4% fat, 5% lactose (milk sugar), and 3 to 4% protein. Those passengers fundamentally alter what the liquid does to your bake.

The fat in milk coats gluten strands. Just as butter shortens gluten (covered in Chapter 4), the fat in milk physically interrupts the gluten network as it develops. Shorter, less-developed gluten means a more tender crumb. The fat lubricates the crumb structure at a molecular level.

Lactose is the sugar naturally present in milk, and it participates eagerly in the Maillard browning reaction. Milk's lactose reacts with the amino acids in the proteins of flour and eggs at high heat, producing hundreds of flavored, browned compounds. That deeper golden-brown top on your milk cake is lactose at work. Milk-based batters can over-brown at high temperatures, so you may need to tent them with foil toward the end of baking.

Milk contains two main protein families: casein (about 80% of milk's total protein, relatively heat-stable) and whey proteins (about 20%, sensitive to heat). This heat sensitivity is why some older bread recipes call for scalded milk, and we will return to that.

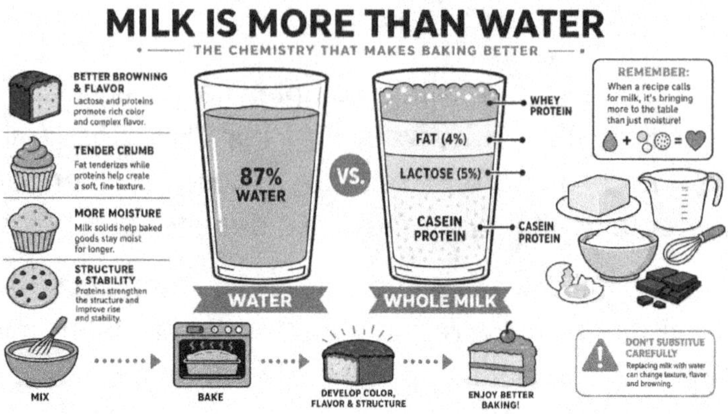

Diagram 6.4: What Milk Carries Compared to Water

Buttermilk: the baking superstar

Traditional buttermilk was the thin, tangy liquid left after cream was churned into butter. Most commercial buttermilk today is cultured: lactic acid bacteria ferment the milk's lactose into lactic acid. The result is thick, tangy, and mildly acidic, with a pH around 4.5. That acidity is not just flavor. It is chemistry.

Baking soda is pure sodium bicarbonate, a base. It needs an acid to activate. When baking soda meets an acid, carbon dioxide bubbles form immediately, giving your batter a lift that is locked in during baking. Buttermilk's acidity provides that partner acid in abundance. The reaction is faster and more powerful than what you get from baking powder alone. A buttermilk biscuit dough must be worked quickly and baked promptly. CO2 production starts the moment buttermilk touches baking soda, and if you dawdle, you lose lift before the oven can set the structure.

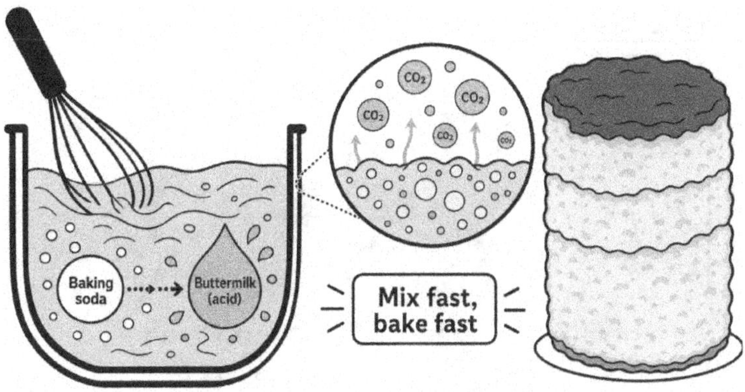

Diagram 6.5: Buttermilk and Baking Soda Reacting

Yogurt, sour cream, and creme fraiche

These three dairy products share a family resemblance: all are thickened by bacterial cultures, all are acidic, and all carry fat. They differ in fat content, thickness, and acidity, and those differences translate directly into baking behavior.

Plain yogurt runs about 3.5% fat for whole-milk versions and has a pH around 4.5, similar to buttermilk. Greek yogurt, strained to remove whey, is thicker and higher in protein. In baking, yogurt activates baking soda, adds a mild tang, and tenderizes gluten. You can substitute it for buttermilk in most recipes by thinning it slightly with water or milk.

Diagram 6.6: The Cultured Dairy Family

Cream: the highest-fat dairy

Heavy cream typically contains 36% or more fat by weight, and that concentration changes almost everything about what it can do. Milk, at 4% fat, can only tenderize slightly. Cream, at 36% fat, can be whipped into a stable foam, transformed into butter, or used to build an impossibly rich custard.

Why can cream be whipped, but milk cannot? When you beat cream, you force air into the liquid. The fat globules partially coalesce around the air bubbles, forming a semi-solid network that holds the air in place. Below about 30% fat, there are not enough fat globules to build that network, and the foam collapses. That is why half-and-half (about 12% fat) cannot be whipped to peaks.

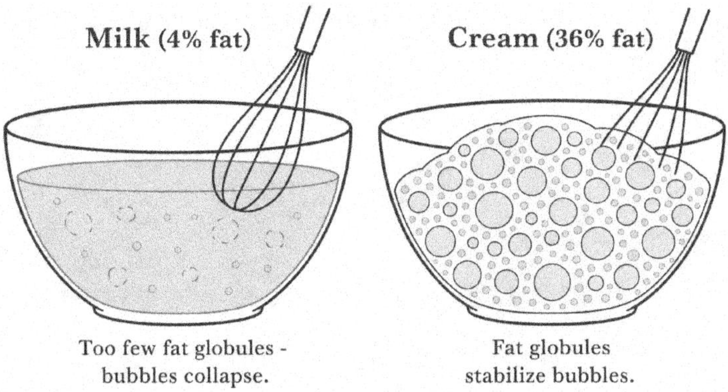

Diagram 6.7: Why Cream Whips and Milk Do Not

Yeast and water temperature

If you bake with yeast, the water temperature is a critical variable you can control directly. Understanding its thermal window turns a vague instruction like "use warm water" into a precise, repeatable skill.

Active dry yeast blooms most reliably in water between 105F and 115F. At those temperatures, the yeast cells rehydrate quickly, wake from dormancy, and begin producing carbon dioxide. The water feels warm but comfortable on your wrist. You may notice the yeast starting to froth within five to ten minutes. That froth confirms the yeast is alive before you commit it to a full batch of dough.

Below about 95F, yeast is sluggish. It will still work, but slowly. A dough made with cool water will rise, but it may take two or three times as long. Cold-proofing a bread dough overnight is a deliberate technique that slows yeast activity and allows more complex flavors to develop. Book 3's fermentation chapter covers this in detail.

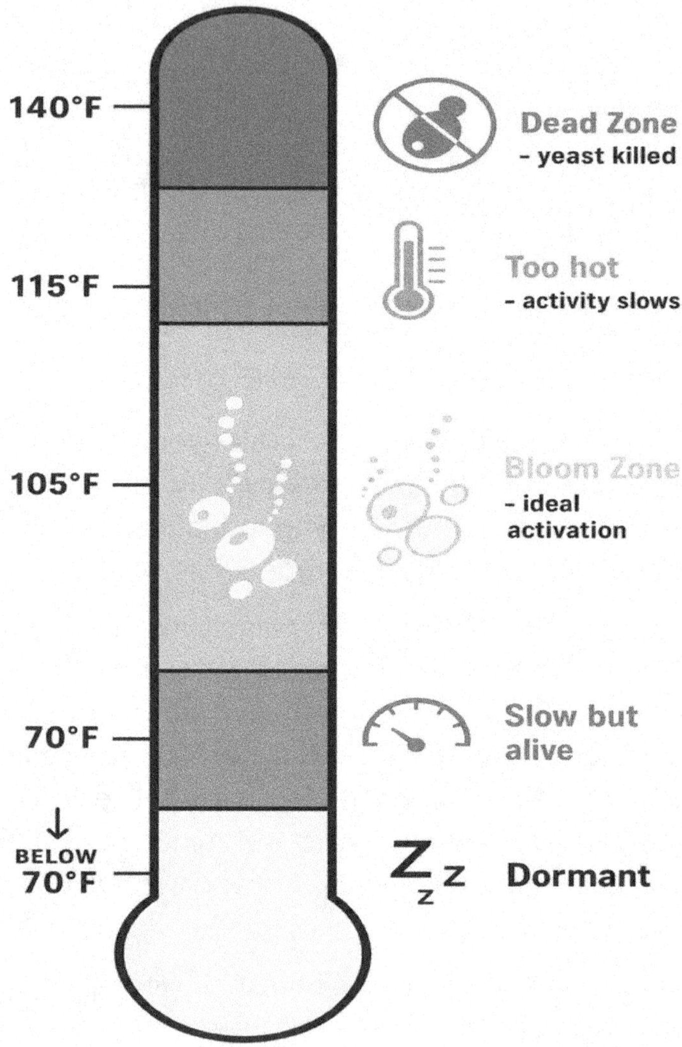

Diagram 6.8: Yeast Temperature Danger Zones

Scalded milk: why old recipes do it

If you have spent any time with older bread recipes, particularly brioche or American sandwich loaf recipes from mid-twentieth-century cookbooks, you have encountered the instruction to scald milk. The recipe tells you to heat the milk to about 180F, watch for tiny bubbles at the edge, then cool it before adding it to the yeast or dough. Many modern bakers skip this step. So what was scalding actually doing?

The answer involves whey proteins. Milk's whey proteins, specifically beta-lactoglobulin, interfere with gluten development when left in their native, undenatured state. In an unscalded milk dough, these proteins bind to gluten strands, disrupting the network and producing weaker, less springy gluten and a denser, faster-staling bread. When you heat milk to around 170F to 180F, you denature those whey proteins. Denaturation means the protein's molecular structure unfolds and loses its normal shape. Once denatured, beta-lactoglobulin no longer interferes with gluten development. The bread made with scalded milk had a stronger gluten network, better rise, and longer shelf life. Scalding was not a ritual. It was applied protein chemistry.

Diagram 6.9: Scalding Milk - Protein Change

How dairy browns: Maillard and lactose

Browning deserves its own space because it affects every dairy product you bake with and explains color differences you may have noticed without understanding why.

The Maillard reaction is a chemical process between amino acids and reducing sugars at high temperatures. We covered it in the heat chapter of Book 1. Lactose, the sugar found in milk and all standard dairy products, is a reducing sugar that eagerly participates in Maillard reactions with the amino acids in flour and dairy proteins. This reaction produces the dozens of browning compounds responsible for the golden crust of milk bread, the deep amber of browned butter, and the rich mahogany crust of a brioche.

Plain water contains no sugars and no amino acids. It does not Maillard-brown at all. A bread made entirely with water browns only through caramelization of the flour's own small residual sugars, a slower and milder process. Add milk, and you add lactose, which dramatically accelerates browning. This is why milk bread browns faster than water bread at the same oven temperature.

Diagram 6.10: Lactose Browning Compared to Water

Emulsions and dairy: a brief connection

If you baked through Book 1's chapter on emulsions, you will recognize some of what milk and cream do structurally as emulsion chemistry. Milk is itself a natural emulsion: fat globules suspended in water, kept from separating by a coating of emulsifying proteins and phospholipids. Cream is a concentrated version of the same emulsion.

Dairy's emulsifying proteins, especially caseins, are part of why milk integrates into batters more smoothly than plain oil. When you add vegetable oil, you need an external emulsifier (typically egg yolk lecithin) to keep it from separating. When you add milk, the natural caseins and phospholipids distribute the fat throughout the batter without extra help.

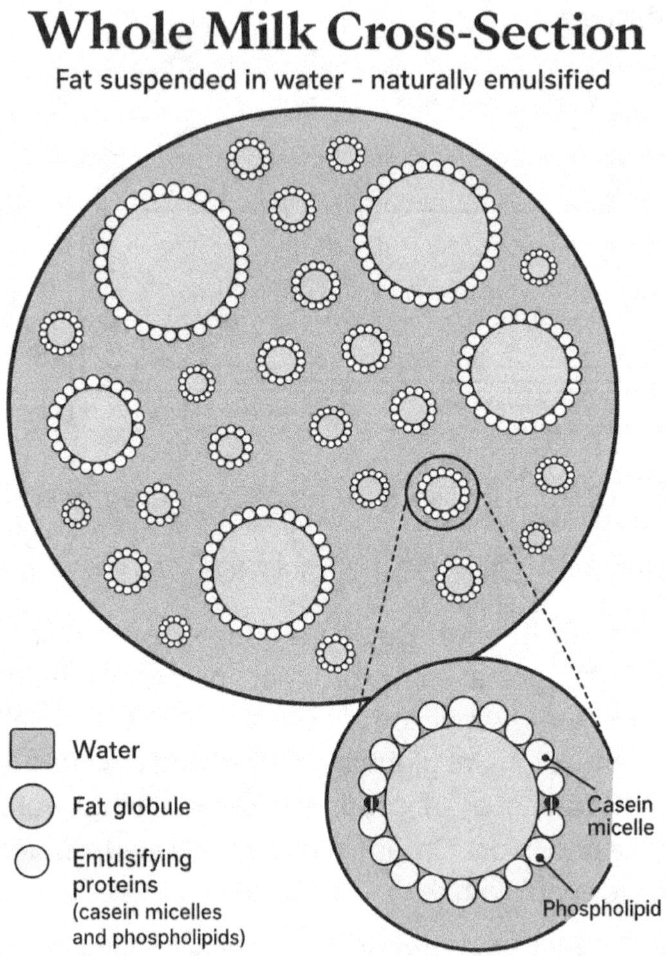

Diagram 6.11: Milk as a Natural Emulsion

Substituting dairy: practical strategies

By now, the principles behind each dairy product are clear enough that substitutions become straightforward decisions rather than guesses.

Replacing whole milk with plant-based milks: oat milk, almond milk, and soy milk all contain significantly less fat than whole milk, though soy milk is the closest in protein content. Oat milk is slightly thick and mildly sweet, with a low potential for Maillard browning. Almond milk is thin, low in fat and protein, mostly water. In most cake and muffin recipes, these plant milks work as one-to-one substitutes for liquid volume, but the fat contribution of whole milk is missing. Adding a small amount of neutral oil compensates for the fat in recipes where tenderness is the primary goal.

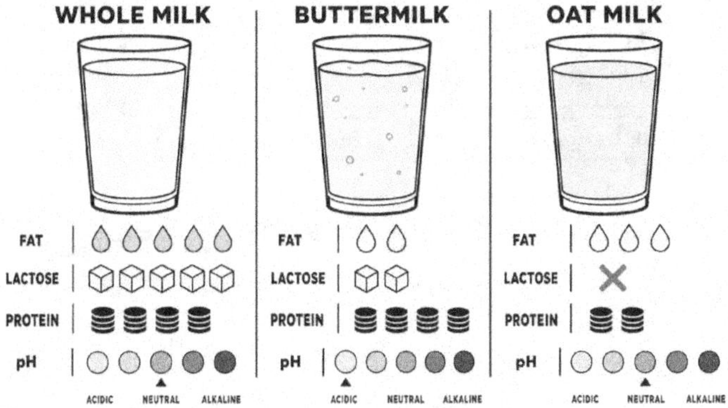

Diagram 6.12: Dairy Swaps and What Changes

The try-this: milk versus buttermilk pancakes

Make two batches of a simple pancake batter at the same time. In one bowl, use whole milk. In the other, use an equal volume of buttermilk or your vinegar-in-milk substitute. Add a small pinch of baking soda to the buttermilk batter, but not to the milk batter. Everything else: identical.

Diagram 6.13: The Pancake Experiment Setup

Temperature, time, and dairy's perishability

Dairy products are alive in ways that water is not. Milk and cream contain proteins that denature with heat, fats that go rancid over time, and, in the case of cultured dairy, active bacteria that continue fermenting even in your refrigerator.

Cold dairy slows the biochemical activity that would otherwise change its character. Buttermilk and yogurt left at room temperature for several hours will be more acidic than refrigerator-cold versions because bacteria continue to produce lactic acid. In a recipe sensitive to acidity, such as a delicate layer cake that relies on a precise baking soda ratio, this shift can affect the rise and texture.

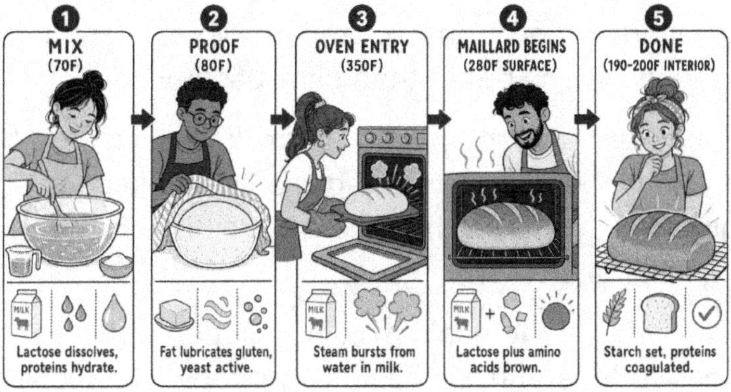

Diagram 6.14: Dairy in the Bake - Temperature Timeline

What you can now do

• Calculate the hydration of any bread recipe by dividing water weight by flour weight and multiplying by 100, then use that number to predict whether a dough will be stiff and shapeable (60%) or wet and open-crumbed (80%+).

• Swap milk for water in a recipe, knowing you are adding fat for tenderness, lactose for faster browning, and protein that will interact with gluten and heat.

Watch the oven time more carefully and expect a richer, softer result.

• Use buttermilk with confidence: pair it with baking soda rather than relying solely on baking powder, work the batter quickly before the CO2 escapes, and expect a taller, more tender result than you get with plain milk.

• Understand why your whipped cream succeeds or fails based on fat content (minimum 35%), temperature (very cold), and pasteurization level (standard pasteurized whips better than ultra-pasteurized).

• Make an acceptable buttermilk substitute by acidifying whole milk with vinegar or lemon juice, or by thinning plain yogurt with a little water: the acidity is the key functional element.

• Decide whether to scald milk in a modern recipe with confidence: pasteurized milk has already had most of its whey protein interference reduced, so scalding is optional except when using raw milk.

• Recognize why milk-based baked goods brown faster than water-based ones at the same oven temperature, and adjust your rack position or bake time proactively rather than pulling out a darker loaf than you expected.

7 The Oven, Heat, Steam, and Oven Spring

You slide a shaped loaf into the oven and close the door. For a moment, nothing seems to happen. Then, slowly and then all at once, the dough lifts. It rises past the edge of the pan, splits open along a score, and nearly doubles in size before your eyes. Ten minutes in, the rise slows. Twenty minutes in, the crust starts to color. Forty minutes in, the loaf sounds hollow when you knock the bottom. You did almost nothing. The oven did everything.

Most bakers treat the oven as a black box. Set the dial, wait. But the oven is a machine with moving parts: cycling heating elements, circulating air, radiating walls, and a floor that acts like a stovetop. Three distinct forms of heat hit your dough simultaneously, each acting differently. Add steam, and you get a fourth player. Add the wrong pan, and you undo an hour of careful proofing.

This chapter focuses on the oven as the final active ingredient. We will look at how heat travels in three modes, why your dial almost certainly lies, what oven spring really is and when it stops, how steam changes the first act of baking, why convection fans speed some things up while wrecking others, and how pan color and thermal mass shape the bottom of everything you bake. If Book 1 taught you about heat as flavor, this chapter is about heat as structure: the engine that inflates, sets, and finishes every baked thing you make.

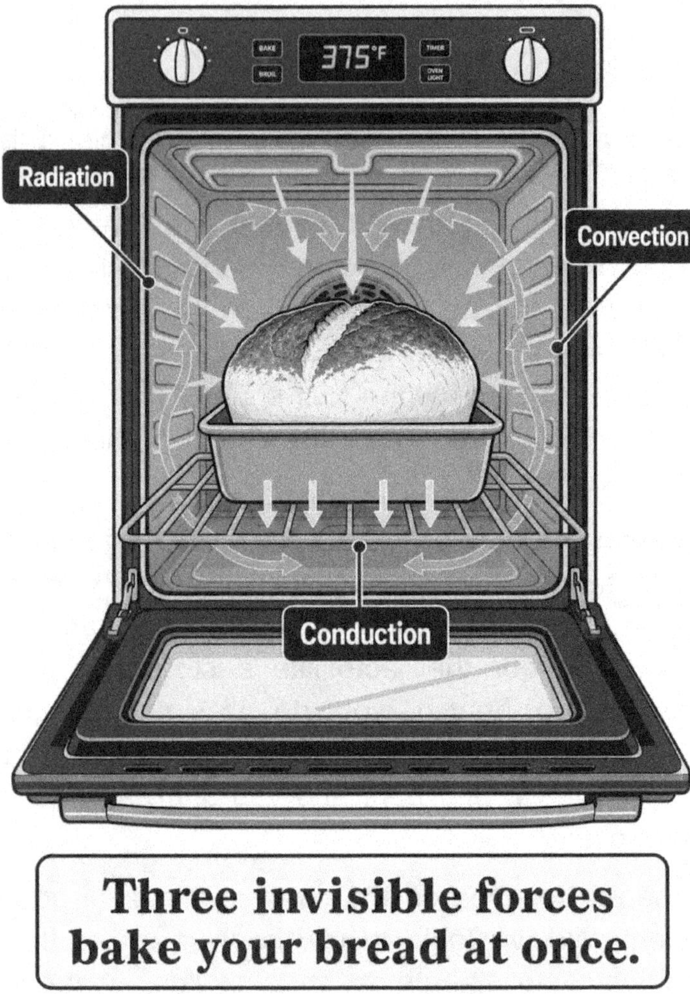

Diagram 7.1: The Oven as a Three-Mode Heat Machine

How Heat Actually Travels in an Oven

Heat does not just "fill up" a space the way water fills a cup. It travels by three distinct mechanisms, and every oven uses all three at once.

The first mode is radiation. Every object warmer than its surroundings emits infrared energy. The glowing heating element at the bottom of your oven does this intensely, and the walls, floor, and ceiling also radiate heat once they are hot. Radiant heat travels in straight lines and does not require air to carry it. That is why a baking stone under a pizza chars the bottom crust before the oven air has had much time to warm it. Radiation scales steeply with temperature: a 50°F increase yields a dramatically larger burst of radiant energy than a proportional increase would.

The second mode is convection. Hot air rises, cool air sinks, and that circulation carries heat from the elements to the food. In a conventional oven, this happens slowly and unevenly, which is why oven hot spots exist and why rotating your pan matters. Convection also moves moisture: hot air picks up water vapor from your dough and vents it, meaning the air near the vents is drier than the air near the center.

The third mode is conduction. Anywhere solid materials touch, heat flows from the warmer object to the cooler one. The oven rack conducts heat to the pan; the pan conducts heat to the dough. The bottom crust gets this direct conductive heat first and fastest, which is why the bottom of a loaf can overbrown before the inside is fully cooked if the pan is too dark or too thin.

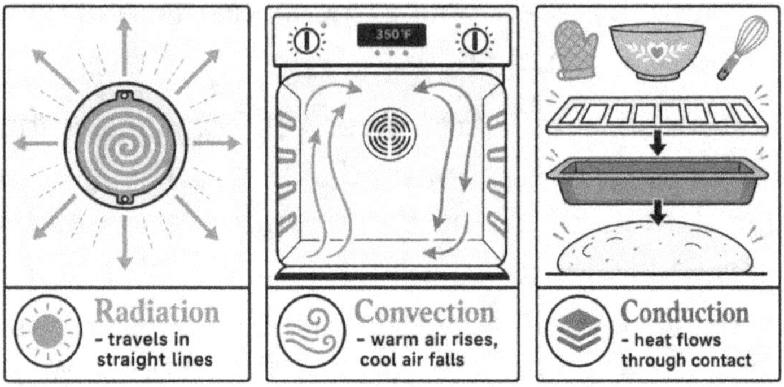

Diagram 7.2: Radiation, Convection, Conduction Close-Up

Your Oven Dial Almost Certainly Lies

The dial on your oven was calibrated at the factory under controlled conditions. By the time the oven has been installed, cycled thousands of times, and lived through years of use, the relationship between the dial and the actual air temperature in the center of the cavity has often drifted by 25F, sometimes 50F. Consumer testing consistently finds that most home ovens run more than 25°F off their stated temperature at higher settings.

This is not a defect. The oven's thermostat measures temperature at one point and switches the element on and off to hold that reading steady. But the element cycles: it blasts heat, the thermostat shuts it off, the temperature drifts down, the element kicks on again.

The actual oven temperature swings by roughly 25°F around the set point at all times. If the thermostat is also miscalibrated, that arc rides on the wrong center.

The fix is simple: buy an oven thermometer. Hang it on the center rack. Leave it in during preheating, and wait an additional 10 minutes after the "ready" beep before trusting the reading, because the first beep usually comes before the walls have fully absorbed heat. Compare the reading to the dial. If your dial reads 375°F and the thermometer reads 350°F, add 25°F to compensate for each bake.

Diagram 7.3: The Oven Thermometer Check

The Four Phases of Baking

When you put dough into a hot oven, it does not do one single thing. It does four things in sequence, each controlled by a different temperature threshold. Understanding these phases is the key to reading a bake in progress.

Phase one is the initial heating phase: the dough warms from room temperature to roughly 100F to 120F inside. The outside heats faster than the inside; the surface begins to dry slightly, and if steam is present, it stays moist and supple during this window.

Phase two is the oven spring. Between roughly 100F and 138F inside the dough, three things happen nearly simultaneously. Yeast goes into overdrive, producing CO_2 faster than at room temperature, in a final burst before the fatal temperature is reached. Gases trapped in the dough expand rapidly as they warm: gas volume rises with temperature, so even without new gas production, existing gas takes up more space. Yeast enzymes also slightly relax the gluten, making the dough extensible enough to stretch around those expanding pockets. A well-fermented loaf can expand 60 to 80 percent in volume if conditions are right.

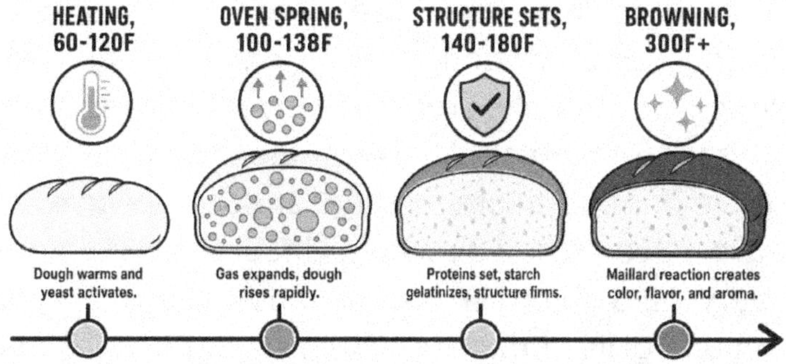

Diagram 7.4: The Four Phases of Baking

Diagram 7.5: Temperature Milestones Inside the Loaf

Oven Spring: What It Is and How to Maximize It

Oven spring is the baker's most visible reward and the most misunderstood event in baking. New bakers think it means the dough "baked well." Experienced bakers know it is the culmination of everything that happened before the oven: proper fermentation, proper hydration, good shaping, and a correct final proof. The oven only creates the conditions for the spring to happen.

The conditions that maximize oven spring are: an adequately proofed dough with enough gas cells developed, a strongly preheated oven, steam in the first 5 to 15 minutes to keep the crust soft and extensible, and a score on the surface to give the expansion a controlled direction. Underproofed doughs give very little oven spring because there are not enough gas cells to expand. Overproofed doughs also show poor oven spring because the gluten network has been stretched to its limit and is too weak to support the expanded pockets; they collapse instead of holding their shape.

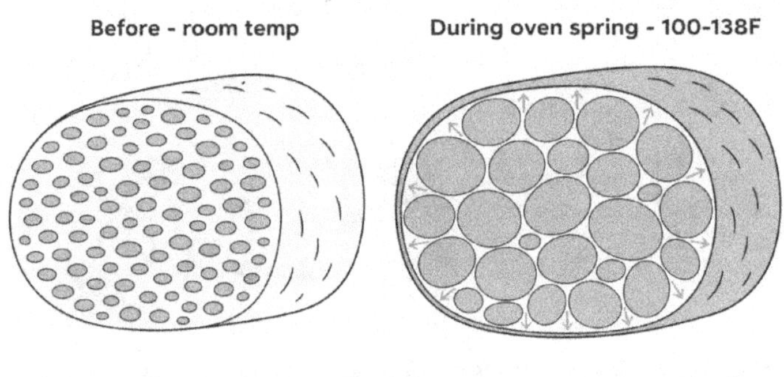

Diagram 7.6: Oven Spring Cross-Section: Gas Cells Expanding

Steam: Bread's Secret Weapon

Steam is the most powerful variable in bread baking that home bakers rarely control deliberately. Professional bakeries invest heavily in steam injection systems: a commercial deck oven can flood its chamber in seconds, hold steam for a precise number of minutes, then vent it completely. Home bakers can achieve the same effect with a little creativity.

In the first 5 to 15 minutes of baking, the dough's surface is hot but still moist. If the oven air is dry, the surface dries quickly, forming a stiff film. If the air is saturated with steam, drying is dramatically slowed. The surface stays moist and supple, the gluten stays extensible, and the dough can keep expanding as long as internal gas pressure keeps pushing.

Steam also affects surface starch. When moist hot steam condenses on the cool surface of the dough, it deposits a thin film of gelatinized starch. That layer, when it later dries in the second phase of baking, becomes the thin crackly exterior that shatters when you break a good baguette. Without steam, the crust is matte and thick. With steam, it is shiny and crackly.

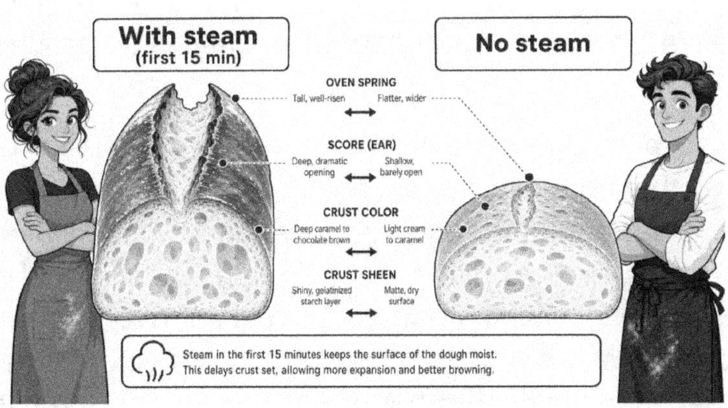

Diagram 7.7: Steam vs. No-Steam Crust Comparison

Diagram 7.8: Three Home Steam Methods

Convection vs. Conventional: Choosing Your Mode

The difference between convection and conventional baking is one fan, but that fan changes a great deal. In a conventional oven, air moves slowly and naturally. In a convection oven, a fan actively circulates air around the cavity.

Forced airflow strips away the thin boundary layer of still air that normally sits just against the surface of your food. That boundary layer acts as insulation, slowing heat transfer. Removing it accelerates heat transfer substantially, which is why convection ovens bake faster: more heat reaches the surface per unit of time, not because the oven is hotter, but because the heat is delivered more efficiently.

The forced airflow also accelerates evaporation. Moving air constantly removes moisture from the surface and carries it to the oven vents. This is great for crispy-skinned chicken and deeply browned cookies. It is not great for a cake that needs to stay moist, or for bread during oven spring, when you want to preserve surface moisture.

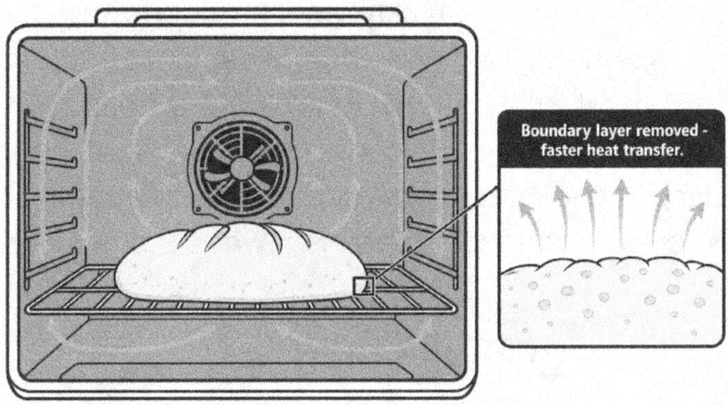

Diagram 7.9: Convection Fan Air Paths

Pan Color, Material, and Thermal Mass

Pan choice is one of the most overlooked variables in home baking. The pan you choose actively controls how much heat reaches the bottom of what you are baking, how quickly it arrives, and how evenly it is distributed.

Color is the most immediate variable. Dark pans absorb more radiant heat and transfer it to the bottom of the food faster. Cookies on a dark baking sheet have darker, crispier bottoms than cookies on a light sheet at the same temperature. Recipes written for light pans will overbrown when baked in dark pans. The rule: switch to a darker pan and reduce the oven temperature by about 25F. Switch to a lighter pan and slightly extend the baking time.

Material matters too. Aluminum is the gold standard: it conducts heat quickly and evenly. Stainless steel conducts heat more quickly and produces more hot spots. Glass heats up slowly but holds heat stubbornly, which is why it takes longer to brown the bottom crust than in a metal pan. Silicone conducts heat poorly, producing pale, lightly colored bottoms. For maximum bottom-crust crispiness, metal is always the answer.

Diagram 7.10: Dark Pan vs. Light Pan Bottom Heat

Diagram 7.11: Pan Material Heat Profiles

Baking Stones and Steels: Thermal Mass at Work

A baking stone or baking steel is a thermal battery. Preheat the oven for 45 to 60 minutes to reach a high temperature. When cold dough contacts its surface, that stored heat transfers instantly and powerfully into the dough's bottom crust. The temperature of the stone or steel does not drop dramatically from one loaf to the next because it has so much thermal mass.

Cold dough placed on a preheated steel will see its bottom crust temperature surge almost immediately. The starch in the bottom crust gelatinizes fast. The bottom structure sets while the top of the loaf is still expanding under steam. The result is a loaf that rises more and develops a markedly crispier bottom crust than the same dough baked on a cold sheet pan.

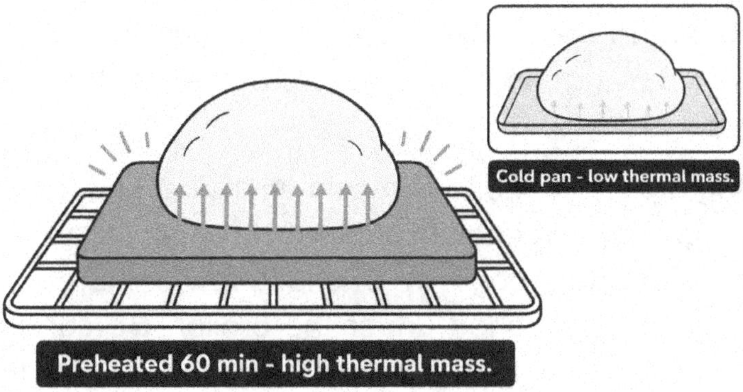

Diagram 7.12: Baking Steel Heat Transfer

Rack Position: The Vertical Geography of Heat

Your oven is not the same temperature from top to bottom, and it is not the same type of heat either. Understanding the vertical distribution of heat helps you place foods in the right zone for what you want them to do.

The bottom rack is closest to the bottom heating element: maximum conduction and radiant heat from below. This is where you put pies when you want the bottom crust cooked through, where your baking stone or steel lives, and where you put the steam pan. The top of a tall loaf will be farthest from the broiler element and may brown slowly from this position.

The center rack is the default for almost everything: farthest from both the bottom element and the broiler, it gets the most balanced combination of all three heat modes. Cakes, quick breads, muffins, and cookie sheets all do best here. If your oven runs hot on the bottom, moving one rack up from the center is a common fix.

The top rack is the zone of maximum radiant heat from above: gratins that need to bubble and crust, bread that needs more top color in the last few minutes, or anything being finished under the broiler. Be careful: items can go from golden to burnt very quickly near the broiler element.

Diagram 7.13: Rack Position Heat Map

The Cost of Opening the Oven Door

Every time you open the oven door, you pay a temperature tax. Room-temperature air rushes in, and rack-level temperature can drop by 50°F or more in seconds. For most items, the cost is extra baking time. But for a few specific situations, opening the door at the wrong moment causes irreversible damage.

Souffles are the classic case. A soufflé is a foam of egg-white-trapped air heated until the proteins set. A rush of cool air chills the surface, the thermal contraction overwhelms the still-fragile protein structure, and the soufflé collapses. The damage is permanent.

Choux pastry is similarly vulnerable in its first phase. If the oven is opened while the shells are still expanding and the exterior is wet, the steam pressure equalizes with room air, the shells deflate, and they will not repuff. Wait until the shells are fully puffed and beginning to color before opening the door.

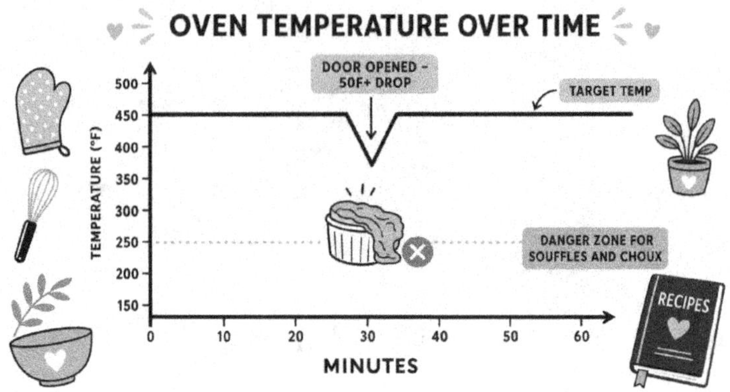

Diagram 7.14: The Door-Opening Temperature Drop

Try This in Your Kitchen

The best way to make oven spring and steam real is to bake two small loaves side by side and change just one variable: steam.

Mix a simple lean dough and divide it into two equal pieces. Shape and proof them side by side. Preheat your oven to 450F with a baking stone or heavy sheet pan on the center rack and a cast-iron skillet on the bottom rack, for at least 45 minutes.

When ready to bake, score both pieces with a sharp blade. Slide one piece onto the hot stone and immediately pour about half a cup of boiling water into the cast-iron skillet. Close the door quickly. Bake the second piece afterward with the skillet dry, or bake both simultaneously with the skillet empty.

After baking, set the two loaves side by side. The steam loaf will be taller, more dramatically split along the score, with a smoother, shinier, thinner crust. Break each one open: the steam loaf will have a slightly more open crumb because extended oven spring allowed more expansion before the structure set. The no-steam loaf will be shorter, matte-surfaced, with a thicker, chewier crust.

Diagram 7.15: The Steam vs. No-Steam Experiment

What You Can Now Do

• **Identify which heat mode is responsible for specific baking outcomes** in your own oven:

overbrowning on the bottom points to conduction from the pan; uneven side-to-side browning points to convection hot spots; a top that browns too fast points to radiant heat from the broiler element.

• **Calibrate your oven with a thermometer** and know your actual baking temperature. Add or subtract the offset for every bake, and map your oven's hot spots with the bread-slice test so you know exactly when and how to rotate.

• **Read a bake in progress through its four phases:** initial heating, oven spring (100F to 138F), structure setting (140F to 180F), and browning (300F and above on the surface). Recognize when each phase occurs and understand why opening the door during the oven spring harms the outcome.

• **Use steam intentionally for lean breads** (baguettes, sourdough, rolls) during the first 15 to 20 minutes via a Dutch oven lid, a steam pan, or a bowl cover. Remove the steam source at the right time to allow the browning phase to proceed cleanly.

• **Choose pans deliberately:** dark pans for more bottom heat and crispiness (reduce oven temperature by 25F), light pans for gentler bottom heat, thick pans for more even heat delivery, and doubled pans to slow bottom heat for delicate cookies.

• **Preheat a baking stone or steel for at least 40 to 45 minutes** to build the thermal mass that delivers professional-quality bottom crust and oven spring to your home-baked breads and pizzas.

• **Switch convection mode on or off by intention:** convection for crispiness and faster browning, conventional for steam retention, delicate custards, souffles, and anything where you want to preserve surface moisture in the early phase of baking.

8 Cookies, the Most Studied Baked Good

You line up six cookies on a sheet of parchment. Same flour, same chips, same chocolate, same kitchen. But each one came from a dough you nudged in a different direction. One cookie lies flat like a lace disc, paper-thin at the edges, almost translucent. Another stands proud and domed, almost like a small muffin that decided it wanted to stay a cookie. The third is crackling crisp, the fourth impossibly soft in the center. The fifth is pale and buttery. The sixth is deep brown, almost caramel-dark, with a chew that pulls back against your teeth. You changed one variable each time. That is it. Same recipe. Six completely different cookies.

This is the most instructive experiment in home baking. Cookies are the perfect laboratory because they are small, they bake fast, they are easy to repeat, and they respond dramatically to every single lever you have been studying since Chapter 1. Gluten formation from your flour decisions, leavening gas from your baking soda, hygroscopic pull from your sugar choice, aeration and fat distribution from your butter temperature, structure from your egg ratio, steam from your liquid, and crust development from your oven temperature all show up together in a two-inch round of dough. No other baked good packs so many variables into such a small space and delivers results in fifteen minutes.

That is why food scientists and pastry school programs study cookies obsessively. The home baker benefits directly. Once you understand the chemistry behind each variable, the perfect cookie stops being someone else's recipe pinned to your refrigerator. It becomes the cookie you engineer for yourself, on purpose, every single time.

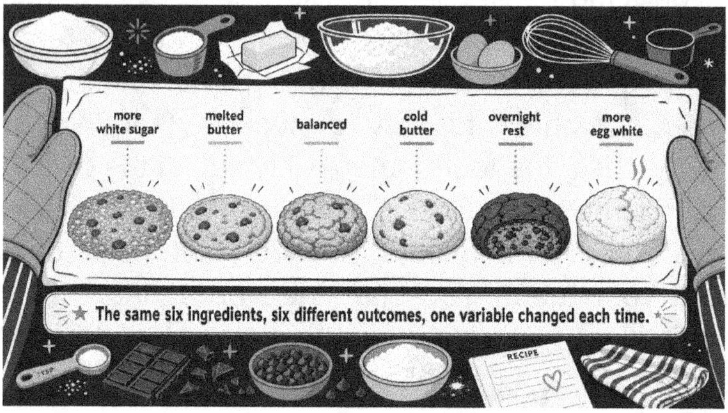

Diagram 8.1: The Six-Cookie Spread Experiment

Why Cookies Are a Perfect Laboratory

Before we go variable by variable, it helps to understand why cookies respond so visibly to small changes. A typical chocolate chip cookie dough contains roughly equal weights of flour and sugar, a significant amount of fat, one or two eggs, a small amount of leavening, and a pinch of salt. That ratio means every ingredient is load-bearing. Unlike bread dough, where extra water or flour can slightly shift the crumb, a cookie has so little margin that a small change in any single component can dramatically affect the final product.

Cookies also bake quickly, which means the oven heat cannot average out your decisions. A bread that bakes for an hour gives your gluten network time to set, your steam time to redistribute, and your crust time to stabilize. A cookie bakes in ten to fourteen minutes. That compressed window locks in whatever your dough brought to the oven, fast. This is why the same small shift in butter temperature that might barely register in a pound cake will produce a radically different cookie.

The other thing cookies reveal beautifully is the competition between spreading and setting. A cookie spreads because the fat melts and the sugars liquefy as the oven heats up. A cookie sets because the eggs coagulate and the starches gelatinize, turning the fluid mass back into a solid. The cookie you end up with is a snapshot of whichever force won the race in your particular oven, with your particular dough temperature, on your particular sheet pan. Every variable you control tips that race one way or the other.

Chapters 1 through 7 each handed you a separate lever. This chapter is where you see all those levers connected. We will walk through each major variable, see exactly how it pushes the spread-versus-set race, then tour the major cookie families to see how each one exploits a different combination of those variables.

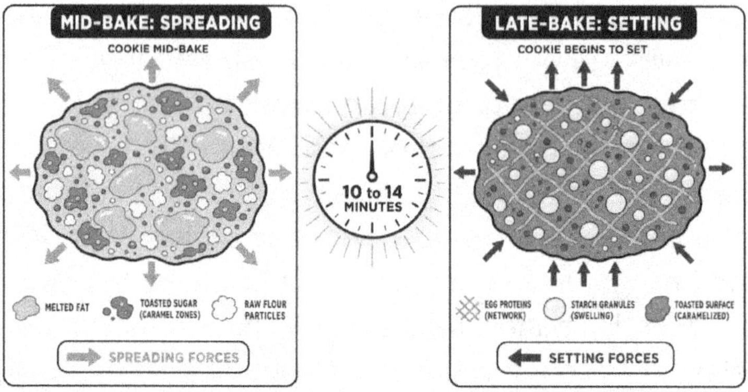

Diagram 8.2: The Spread-vs-Set Race Inside the Oven

The Six Variables, Ranked by Impact

Not every variable matters equally. After watching thousands of bakers run cookie experiments, a clear ranking emerges. The variable that moves the needle most dramatically sits at the top. Work your way down, and the effects get subtler, but they are never zero.

Sugar ratio tops the list, which surprises some bakers who expect butter or flour to dominate. Sugar does several things at once inside a cookie. It dissolves during baking and then recrystallizes as the cookie cools, which determines the final texture more than almost anything else. White granulated sugar recrystallizes into a crunchy, snapping structure. Brown sugar, which contains molasses, holds onto water molecules even after baking because molasses is hygroscopic, meaning it actively attracts moisture from the surrounding air and from the cookie interior. The more brown sugar you use, the more water the finished cookie holds, and the chewier it stays, even days later. A cookie made with all white sugar will be crisp and snap when you break it. A cookie made with all brown sugar will bend before it breaks. Shift the ratio between them, and you slide along that spectrum.

White sugar also spreads the cookie more than brown sugar does. Granulated white sugar dissolves faster in the hot oven and lubricates the dough, allowing it to spread before the egg proteins have time to set the structure. Brown sugar, being slightly coarser and moister, dissolves a little more slowly. All else being equal, more white sugar means a wider, thinner, crispier cookie. More brown sugar means a taller, softer, chewier one.

Butter temperature ranks second, and its effect is profound. This goes back directly to what you learned in Chapter 4 about fat aeration. When you cream cold or room-temperature butter with sugar, the sugar crystals physically cut into the fat, creating millions of tiny air pockets. Those air pockets expand in the oven, puffing the cookie upward and giving it structure. The cookie holds its shape better because it arrived in the oven already full of tiny gas bubbles.

Melted butter changes everything. When butter is liquid before it even touches the sugar, there is no creaming. No air gets trapped. The fat coats the flour particles more completely, slightly suppressing gluten formation by acting as a physical barrier between water and flour proteins. A melted-butter cookie goes into the oven dense, flat, and with less structure to resist spreading. It spreads wide and thin. The edges crisp up. The center stays chewy and dense. Many food scientists argue that the best chocolate chip cookies are made with melted butter, specifically for this reason, not despite it.

Diagram 8.3: Butter Temperature and Cookie Shape

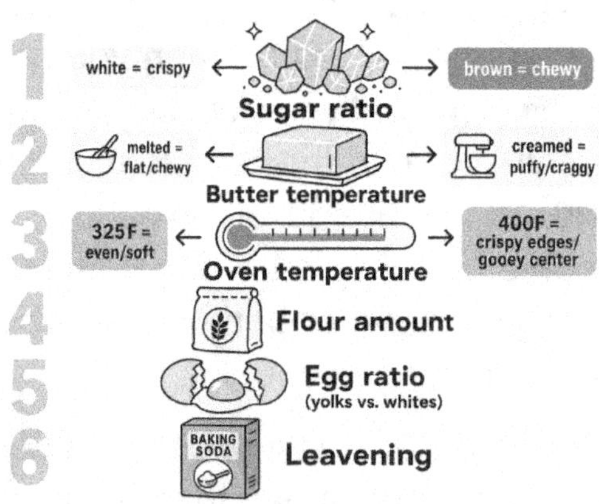

Diagram 8.4: The Six Variables Ranked by Impact

Crispy, Chewy, or Cakey: The Decision Tree

Every cookie outcome you have ever eaten sits somewhere on a triangle with three points: crispy, chewy, and cakey. You can use the chemistry you now understand to navigate toward any point on that triangle deliberately.

A **crispy cookie** wins when you maximize spread and minimize retained moisture. White sugar dominates the blend. Butter is softened but not melted, and the amount is relatively generous, so the cookie spreads thinly. Baking time is longer, and oven temperature is moderate, giving the cookie time to dry out fully. Thin cookies with high white-sugar content are also more prone to caramelization, adding that glassy, amber-crisp edge. Shortbread is at the end of this path: almost pure butter and white sugar, with minimal egg and just enough flour to hold its shape. It dries completely in the oven and snaps cleanly.

A **chewy cookie** wins when you trap moisture and maximize Maillard browning without crisping the full cookie through. Brown sugar dominates. Melted or browned butter goes into the dough. Extra yolks add fat and richness. The oven is hot enough to set the edges fast and brown the bottom, but the center comes out before it dries. The cookie cools on the sheet rather than on a rack, letting carryover heat gently finish the interior. Resting the dough overnight amplifies every one of these effects, as you will see in the next section.

A **cakey cookie** wins when you build structure faster than spread can happen. More flour stiffens the dough. Whole eggs (or extra whites) add structure-forming proteins. Baking powder adds lift. Butter is creamed rather than melted, trapping air that pushes the cookie upward instead of outward. Less sugar also reduces the liquefying spread force. Snickerdoodles and certain soft sugar cookies lean cakey because their ratios move in this direction: more egg, more flour, creamed butter, and the cream of tartar in a snickerdoodle controls pH in a way that gives the cookie that classic velvety bite.

The decision tree is this: ask what you want in the finished cookie, then trace back to the ingredients that produce that outcome. Want crispier? Shift sugar toward white, bake longer at moderate heat. Want chewier? Shift to brown sugar, use melted butter, pull from the oven when the center still looks underdone. Want cakier? Add a bit more flour and an extra whole egg, cream the butter well, and consider adding a small amount of baking powder.

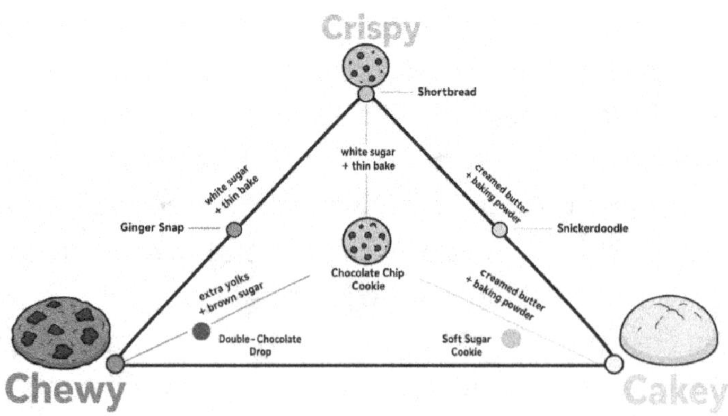

Diagram 8.5: The Crispy-Chewy-Cakey Triangle

The Science of Brown Sugar and Chew

Brown sugar is not just white sugar with coloring added. It is white sugar recombined with molasses, the thick dark syrup left behind when sugarcane juice is boiled down to extract sugar crystals. Light brown sugar contains around 3.5 percent molasses. Dark brown sugar contains about 6.5 percent. That molasses content is what makes brown sugar such a powerful tool in cookie baking.

Molasses is hygroscopic. It pulls water molecules toward itself and holds them tightly. This property means a cookie made with brown sugar will always retain more moisture than one made with the same weight of white sugar, even after baking, even days later. The moisture is not loose water that you can feel or squeeze out. It is water-bound into the molasses compounds, staying in the cookie structure. This bound water is what you experience as chew. The cookie bends rather than snapping because the retained moisture still plasticizes the structure. Leave a brown-sugar cookie out on the counter for two days, and it is still soft. Leave an all-white sugar cookie out for two days, and it turns crisp. Both are doing exactly what their chemistry says they should.

The molasses also contributes to color. Molasses contains compounds that brown dramatically under heat, and those same compounds react with the amino acids in eggs and flour proteins to drive the Maillard reaction at a lower temperature than pure sucrose alone would. A batch of cookies made with dark brown sugar will come out noticeably darker than one made with light brown sugar, even at the same oven temperature and baking time. This is chemistry you can see and taste: the darker color brings with it deeper, more complex flavor notes, slightly bitter and caramel-like, that balance the sweetness.

One underappreciated consequence of brown sugar's molasses content is that it makes the dough slightly more acidic. Molasses is mildly acidic, and its acidity reacts with baking soda to produce carbon dioxide gas, which gives the cookie a bit of lift even with small amounts of leavening. This is why most classic chocolate chip cookie recipes specify brown sugar and baking soda together rather than baking powder: the two are designed to react with each other.

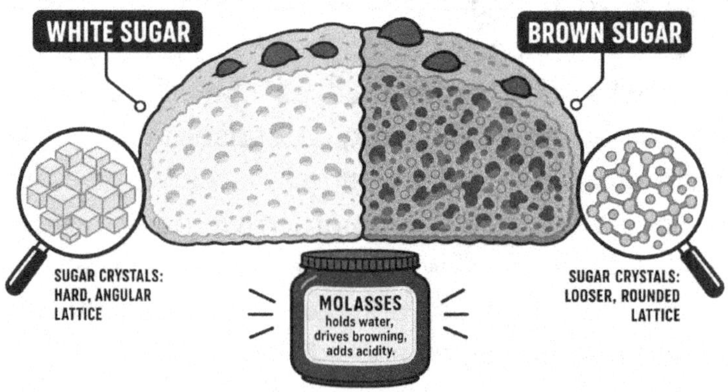

Diagram 8.6: What Molasses Does Inside a Cookie

Butter Temperature and What It Does to Structure

You have already seen how butter temperature ranks second in the variable hierarchy. Now look at the mechanism in detail, because the specific chemistry of cold versus melted butter in cookies explains phenomena that puzzled bakers for generations.

Creaming butter and sugar together, at cool room temperature around 65F, is fundamentally a mechanical process. The sugar crystals have sharp edges, and as the mixer beats them through the fat, those edges physically cut open the fat, creating tiny pockets. Air gets trapped in those pockets. Under a microscope, you would see the creamed butter as a foam: millions of fat-surrounded air bubbles. This is the same principle you learned in Chapter 4 when you saw how butter's fat crystal network determines how well it can hold air.

Those air pockets are critical to cookie structure. When the cookie enters the oven, the air expands as it heats, pushing outward in all directions. This gives the cookie a lift and a slightly domed, craggier surface. The trapped air also slows the flow of the melting fat, because the fat has to travel through a foamy matrix rather than pooling freely. Creamed-butter cookies set more upright.

Melted butter has already broken that fat crystal network before it touches the sugar. There is no foam to create. When you stir melted butter with sugar, the sugar dissolves partially into the butter rather than cutting air pockets into it. The resulting mixture is denser and more liquid. When this dough hits the oven, the fat is already mobile. It spreads quickly across the hot sheet pan before the eggs can coagulate. The cookie flows outward rather than puffing upward.

But here is the twist: melted butter makes a superior chewy cookie in ways that go beyond just spread. The lack of air pockets means the cookie is denser, with a tighter crumb structure. The higher fat-to-air ratio means the cookie tastes richer because every bite contains more fat molecules, without dilution from air. The slightly more complete coating of flour particles with liquid fat also modestly suppresses gluten formation, making the texture more tender. All three effects point in the same direction: the melted-butter cookie is chewy, dense, rich, and tender, which is exactly what most people mean when they say they want a "perfect" chocolate chip cookie.

Brown butter amplifies this further. When you melt butter and keep heating it until the milk solids toast at the bottom of the pan, you drive off about 18 percent of the water that butter normally contains. You also create hundreds of new flavor compounds through the Maillard reaction in those milk solids: nutty, caramel-like, slightly toasted notes that raw or merely melted butter lacks. Brown butter going into a cookie dough delivers more intense flavor, less water (which means less steam and less spread-from-steam during baking), and a richer color. The combination of brown butter plus brown sugar is one of the most flavor-dense combinations in all of cookie baking, and Chapter 4 explains exactly why each element contributes the way it does.

Diagram 8.7: Creamed vs. Melted vs. Brown Butter

Overnight Rest: What Dough Does in the Refrigerator

Baking the dough immediately after mixing, rather than resting it in the refrigerator overnight, might seem like a scheduling convenience rather than a chemical decision. It is one of the most impactful chemistry decisions in cookie baking.

When you finish mixing cookie dough, the flour has just been wetted. The water in your eggs and in any brown sugar or honey has begun to hydrate the starch granules and the proteins in the flour, but that process takes time. Many starch granules are still dry at their cores. The proteins have begun forming gluten strands but have not yet fully organized. The entire dough is, in some sense, still in flux.

Over the next twelve to seventy-two hours in the cold, several things happen. First, hydration continues. Water slowly migrates from the eggs and sugars into the flour granules, hydrating them more completely. Fully hydrated flour starch and protein behave differently from partially hydrated flour: the starch gelatinizes more readily in the oven, resulting in a better set, and the proteins form a more organized gluten network. The result is a cookie that holds its shape better and has a more defined structure.

Second, and perhaps more dramatically, enzymes naturally present in flour go to work. Amylases are enzymes that break down long starch chains into shorter sugar molecules, such as maltose and glucose. These enzymes work slowly at cold temperatures but remain active. Over twelve to twenty-four hours of resting, the amylases break down a measurable fraction of the flour starches into these smaller sugars. This matters enormously for flavor and color. Smaller sugar molecules are more available for Maillard reactions and caramelization in the oven. A rested dough produces a noticeably darker, more deeply browned cookie. Those smaller sugars also taste sweeter and more complex than the raw starch they came from. A rested cookie genuinely tastes sweeter, even if you didn't add a single extra grain of sugar.

Third, some of the water in the dough evaporates slowly, even in the refrigerator, slightly concentrating the flavors. The fat solidifies and redistributes through the dough, creating a more uniform texture.

When you bake the rested dough and the immediately baked dough side by side, the differences are visible before you even taste them. The rested cookie is darker. Its edges are more defined. Its caramel notes are deeper. Its chew is more pronounced. Many professional bakeries rest their cookie doughs for twenty-four to seventy-two hours precisely because the chemistry is undeniably better, even if it requires planning.

Diagram 8.8: What Happens During Overnight Rest

Brown Butter, Brown Sugar, and the Chewy Cookie

Now bring together the three biggest flavor-and-texture multipliers in cookie science and see what happens when they work in combination: brown butter, brown sugar, and a higher ratio of egg yolks to whites.

You have already learned what each does individually. Brown butter contributes toasted flavor compounds and reduced water. Brown sugar contributes hygroscopic moisture retention, acidity for leavening, and deep caramel color. Extra yolks contribute fat-based richness, emulsification, and Maillard-active amino acids that promote further browning.

Now observe what they do together. The reduced water from brown butter and the hygroscopic water-holding from brown sugar work in opposition in an interesting way. The brown butter has less water to contribute to steam, which slows the cookie's early spread because less steam is pushing it outward. But the molasses in brown sugar holds the water present extremely tightly in the finished cookie. So the cookie spreads somewhat less due to reduced steam, then chews more due to locked-in molasses moisture. The extra yolks add enough fat to keep the interior lush and suppressed from over-setting, so the center stays dense and soft even when the edges are fully baked.

The flavor result of this combination is the deepest any home baker can reach without specialty ingredients. The brown-butter Maillard compounds, the molasses caramel compounds, and the yolk-enriched browning compounds create overlapping layers of flavor: nutty, caramel, slightly bitter chocolate-adjacent notes that taste almost like toffee in the crust and fudge in the center. This is why brown butter chocolate chip cookies have become something of a calling card in serious home-baking communities. The chemistry behind them is not mysterious once you understand what each ingredient is doing.

This combination also illustrates a broader principle in cookie baking: the best results often come from stacking several chemical shifts in the same direction rather than relying on a single large change. You could use melted (but not browned) butter and get a chewier cookie. You could use dark brown sugar and get a chewier cookie. You could add extra yolks and get a richer cookie. But doing all three at once amplifies each effect and creates something more than the sum of its parts.

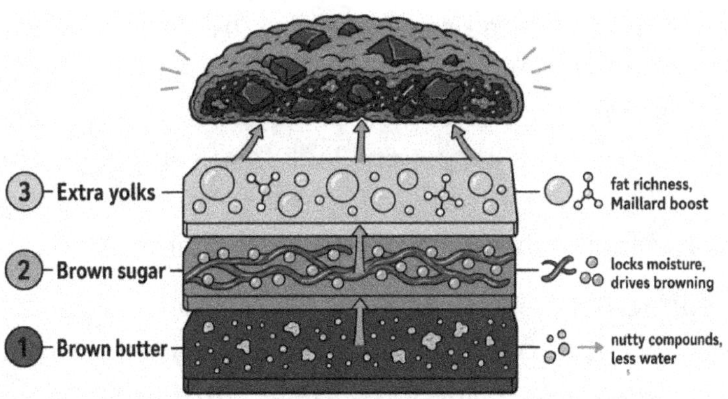

Diagram 8.9: The Triple Stack of Chewy Chemistry

Why Chocolate Chip Cookies Spread and Crack

The classic American chocolate chip cookie is designed to spread, crack, and partially collapse.

These are features, not defects, and they happen through a specific sequence of events in the oven that you can now trace step by step.

The dough enters the oven with butter distributed throughout, either creamed or melted. As the oven temperature climbs, the butter fat melts somewhere between 90F and 110F, well below the 325F to 375F oven temperature most recipes use. The moment the butter liquefies, the dough loses its structural support and begins to spread outward across the sheet pan. This early spread is the first feature: the flat, wide footprint of the finished cookie.

As the cookie spreads and the temperature climbs toward 160F to 180F, the egg proteins begin to coagulate. This is the race you already understand. In most chocolate chip cookie recipes, the sugar and fat quantities are large enough that the spread gets a good head start before the proteins lock things in place. This is not a flaw in the recipe. It is the calibrated balance that produces a flat, wide cookie with crispy edges.

The surface of the cookie dries first. The top is exposed directly to hot oven air, which quickly evaporates surface moisture. As the top surface firms up and dries, the interior remains fluid and slightly expands from heat. The expanding interior pushes against the dried top crust, creating cracks. Those cracks are the visual signature of the classic chocolate chip cookie: the craggly, uneven surface that breaks into shards of crispy crust over a soft interior.

The slight underbaked center is intentional. A chocolate chip cookie removed from the oven while the center still looks glossy and soft will continue to cook for 5 to 10 minutes on the hot sheet pan. The bottom of the pan holds heat well, and that heat conducts upward through the cookie—the egg proteins in the center finish coagulating. The starches finish gelatinizing. By the time the cookie looks fully set on the sheet, it has overshot the ideal. The bakery technique is to pull the cookie when it appears underdone, let carryover heat finish it, and transfer it to a rack only when the cookie can hold its shape without the pan's support.

Diagram 8.10: The Spread and Crack Sequence

Resting Dough: The One-Variable Experiment

Before we tour the cookie families, here is the Try This in Your Kitchen experiment that makes all of this tangible and memorable. It requires only one batch of your usual chocolate chip cookie dough and an overnight wait.

Mix your dough as you normally would. Portion it into balls and separate it into two equal groups. Bake the first group right now, fresh from the mixing bowl. Wrap the second group in plastic wrap and refrigerate it overnight, at least twelve hours. The next day, bake the refrigerated dough at the same temperature and for the same time as you baked the first batch yesterday.

Place one cookie from each batch side by side on your counter. Before you taste them, look. The overnight batch is darker. The color is deeper, more toasted caramel and chocolate brown, compared to the paler golden mustard yellow of the fresh batch. The edges of the rested cookie are more defined. If you pull one apart, the interior of the rested cookie shows a tighter, more chewy crumb. The fresh batch may be slightly softer and more uniform, but also paler and blander-looking.

Now taste them. The rested batch is sweeter, even though no extra sugar was added. That sweetness comes from the amylase enzymes you read about earlier: they have been converting flour starches into smaller, sweeter sugar molecules for the past twelve hours. The flavor is more complex, with caramel and toffee notes that the fresh batch lacks. The chew is more pronounced because the fully hydrated flour structure holds together differently than partially hydrated dough does.

You have just run a controlled food science experiment with a single independent variable: rest time. Every other condition was identical. The result is a real, measurable, tasted difference caused by chemistry you can name and explain. That is what this chapter is about: turning kitchen intuition into kitchen understanding.

Diagram 8.11: Rested vs. Fresh Dough, Side by Side

A Tour of Cookie Styles and Their Chemistry

Every major cookie family in the Western baking tradition is essentially a specific set of variable settings, dialed to produce a particular outcome. Once you understand the variables, you can read any cookie recipe and predict the result before you bake a single one.

Chocolate chip cookies are the balancing act you have studied throughout this chapter. They sit in the center of the crispy-chewy-cakey triangle, tilted slightly toward chewy in most contemporary versions. The defining feature is the chocolate chip itself, which has its own chemistry: chips are formulated with less cocoa butter than eating chocolate, so they hold their shape in the oven rather than melting flat. The chips melt enough to become gooey inside the cookie, but not so much that they spread into a puddle.

Shortbread is the most butter-forward cookie in the canon. The ratio of butter to flour is higher than in any other style, and there is little to no egg. Without eggs, there are almost no proteins to set a structure, so the cookie relies almost entirely on the fat-and-starch matrix to hold together as it bakes. The result is a cookie that is extremely tender, very crumbly, and almost sandy in texture: bakers call this "short" because the fat literally shortens the gluten strands, preventing them from forming a chewy network. This is the same "shortening" action you learned about in Chapter 4, taken to its logical extreme. Shortbread is also typically baked at lower temperatures, around 300F to 325F, to prevent browning. The goal is a pale, delicate, butter-forward cookie that melts in the mouth rather than one that chews.

Snickerdoodles are defined by cream of tartar, an acid that does two things in this cookie. First, it reacts with baking soda to produce carbon dioxide leavening, creating a slightly cakey lift. Second, and more subtly, the acid lowers the dough's pH, which slows the Maillard reaction and keeps the cookie pale and soft rather than dark and crispy. The velvety, almost tangy character of a classic snickerdoodle comes directly from that acidic pH environment. The cookie is rolled in cinnamon sugar before baking. That coating becomes slightly gritty and sparkling as the sugar caramelizes in the oven, creating the signature crackly exterior over a soft interior.

Sugar cookies are close relatives of shortbread, but with more egg and more sugar, which makes them sturdier for rolling and cutting. The classic sugar cookie is made for shape retention: you cut out stars, Christmas trees, and snowflakes, and you want them to look like stars, Christmas trees, and snowflakes after baking. The flour is higher relative to fat than in shortbread, which builds more gluten structure. The egg proteins set quickly and prevent excessive spread. The baking temperature is moderate, around 350°F, and the goal is a full set with minimal browning so the royal icing can sit on a pale background. The chemistry here is deliberately conservative: you are asking the cookie to hold its edges and resist all the spread forces that other styles encourage.

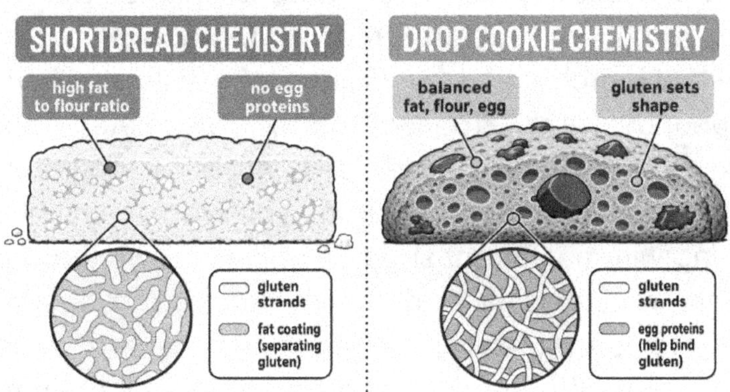

Diagram 8.12: Shortbread Chemistry vs. Drop Cookie Chemistry

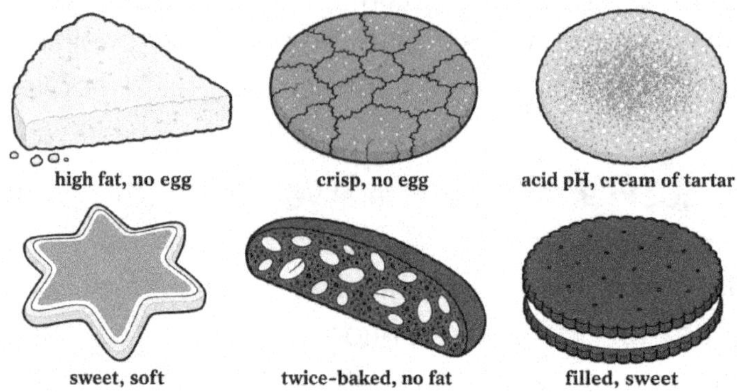

Diagram 8.13: A Cookie Style Gallery

The Maillard Reaction and Caramelization in Cookies

You have seen these two browning reactions in earlier chapters. Still, cookies give you a controlled environment to watch both in action and distinguish between them because you can isolate the variables.

Maillard browning happens when amino acids and reducing sugars react at temperatures above roughly 280F. It produces hundreds of distinct flavor compounds: nutty, roasty, meaty, chocolatey, and toasty. Every one of those flavor notes is a different chemical compound formed from a different amino acid reacting with a different sugar. Cookies brown through Maillard reactions primarily in the crust and bottom, where temperatures are highest and where direct contact with the hot pan drives the reaction. Eggs contribute amino acids. Flour proteins contribute amino acids. Both are necessary for Maillard browning. A cookie with no protein, hypothetically, would not brown through Maillard reactions even if you baked it forever.

Caramelization is a separate reaction involving only sugars, no protein required. It begins at higher temperatures than the Maillard reaction: sucrose starts to caramelize around 320F to 340F, producing the characteristic caramel flavor and an amber-to-dark brown color. In a cookie, caramelization is most obvious at the very edge of a thin, spread-out cookie, where temperatures are highest, and moisture has evaporated most fully. The glassy, amber, almost see-through edge of a thin, crispy cookie is pure caramelization. Break that edge, and it snaps cleanly. Put it on your tongue, and it melts into a sweet, slightly bitter, complex caramel note.

Brown sugar promotes both reactions. Its molasses content contributes to the formation of reducing sugars (glucose and fructose from partial inversion), which are more reactive than sucrose in Maillard reactions. Its acidity shifts pH in a direction that favors Maillard browning. And the molasses compounds themselves darken under heat. This is why brown sugar cookies brown faster and at lower oven temperatures than white sugar cookies, and why the flavor of a dark brown sugar cookie is so layered.

The bottom of the cookie matters enormously. A dark, thick sheet pan absorbs heat faster and reaches higher temperatures than a light-colored, thin pan. This accelerates both Maillard and caramelization reactions at the cookie's base. If you have ever had a cookie that is perfectly done on top but too dark on the bottom, the pan conductivity is the direct cause. A light-colored, insulated, or doubled sheet pan slows heat transfer to the bottom, giving the cookie's interior more time to set before the bottom overcooks. This is why professional pastry kitchens use heavy, light-colored aluminum pans rather than dark steel ones for most cookies.

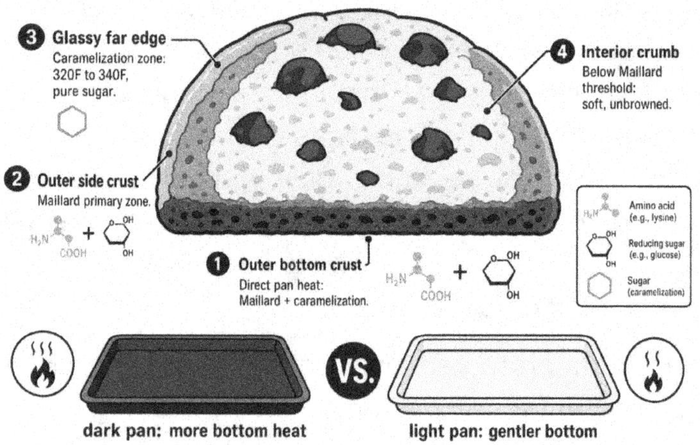

Diagram 8.14: Maillard vs. Caramelization in Cookies

Gluten's Quiet Role in Cookies

Gluten plays a supporting role in cookies rather than the lead role it plays in bread, but it still significantly shapes the outcome. Cookie dough contains far less water than bread dough, so gluten development is limited from the start. This is by design. Too much gluten in a cookie creates a tough, leathery, chewy-in-the-wrong-way texture rather than the pleasant chew of a well-made chocolate chip cookie.

The main strategies for controlling gluten in cookies are fat, minimal mixing, and sugar. Fat coats the proteins in flour, physically preventing them from hydrating and bonding with each other. This is the same mechanism you learned about in Chapter 4, and it is why high-fat cookies like shortbread are so tender: there is so much fat that gluten can barely form at all. Mixing time matters too: unlike bread dough, which you knead to develop gluten, cookie dough is typically mixed only until just combined. The moment you see a homogeneous dough, you stop. Each additional revolution of the mixer forms a few more gluten strands.

Sugar competes with flour for water. Because sugar is highly water-soluble, it binds a significant portion of the available water before it ever reaches the proteins in the flour. Less water reaching the proteins in the flour means less gluten formation. This is one of the reasons very sweet cookie doughs are often tender: the high sugar concentration actively prevents gluten development by monopolizing the water supply.

Flour protein content is also a choice. All-purpose flour, at around 10 to 12 percent protein, is the standard for most cookies. Cake flour, at around 8 to 9 percent protein, produces a more tender, delicate cookie. Bread flour, at 12 to 14 percent, produces more chew in a cookie. Some bakers deliberately use bread flour in chewy chocolate chip cookies to get a slightly more substantial, stretchy chew. If you have read Chapter 1, you know exactly why: the higher protein content provides more glutenin and gliadin to form the gluten network that gives that pull-and-stretch sensation when you tear the cookie.

Diagram 8.15: Gluten Control in Cookie Dough

Eggs, Steam, and the Cakey Possibility

Eggs serve two opposing roles in cookies simultaneously, and the ratio of whites to yolks determines which role dominates. You met this concept in Chapter 5, and now you see it concentrated into a small piece of dough.

The whites of an egg are roughly 90 percent water and 10 percent protein. That water turns to steam in the oven, which puffs the cookie upward the same way steam puffs pastry. Those proteins set into a firm network at around 160F to 180F, which locks the puffed structure in place. A cookie with a relatively high proportion of egg whites will be cakey, lighter, and higher in profile. It will also be slightly less rich, because white contributes almost no fat.

The yolks of an egg are roughly 50 percent water, 50 percent fat, plus protein and emulsifiers. The fat slightly suppresses structure formation, the emulsifiers help fat and water stay blended in the dough, and the protein still contributes to the set. But the dominant yolk effect is richness and density. Yolk fat coats proteins similarly to butter fat, moderating the protein network and preventing it from over-setting. The result is a denser, richer, chewier cookie without the cakey lift that whites provide.

This is why so many professional recipes for chewy cookies call for one whole egg plus one extra yolk. The whole egg provides enough white to give you some structure and moisture for browning, while the extra yolk tips the fat-to-water ratio of the egg component decisively toward richness and chew. Drop the extra yolk and use the white instead, and the same recipe makes a noticeably lighter, cakier cookie. Keep the white and lose the extra yolk, and the cookie gets taller and less rich. The egg is a tunable dial.

The steam from egg whites also interacts importantly with the surface crust. Steam escaping from the interior of the cookie as it bakes pushes up against the partially dried top surface, creating the characteristic puffy dome and the small cracks you see in cakey cookies. In a chewy cookie made with extra yolks and less white, there is less steam, less internal pressure, and the top surface sets flatter without the dome and without dramatic cracking.

Diagram 8.16: The Egg Ratio Dial

What You Can Now Do

You have walked through the complete science of cookies, from the molecular behavior of sugar and butter to the enzyme activity in refrigerated dough, from the Maillard and caramelization zones in a baked crust to the competing roles of egg whites and yolks. Cookie baking is no longer a matter of following someone else's formula and hoping for the best. Here is what you can now do:

• **Engineer your cookie's texture from the inside out.** Shift the sugar ratio toward brown for chewier results, toward white for crispier ones. Use melted butter for a flat, dense, chewy cookie. Use creamed cold butter for a puffier, craggier one. Every change has a predictable chemical cause.

• **Control browning deliberately.** High-brown-sugar doughs brown faster and at lower temperatures. Dark sheet pans accelerate Maillard and caramelization at the base. Light, insulated pans slow it. You choose the outcome before the cookie enters the oven.

• **Rest your dough overnight and taste the difference yourself.** Amylase enzymes convert starch in flour into sweeter, more reactive sugars over 12 to 24 hours. The rested cookie is demonstrably darker, sweeter, and more complex. You can explain every part of why.

• **Navigate the spread-versus-set race.** A hot oven (400F) sets the edges fast and leaves the center gooey. A cooler oven (325°F) gives the cookie time to spread evenly before it sets. You now know which to choose for the result you want, and why.

• **Read any cookie recipe and predict the outcome.** Look at the sugar split, the butter state, the egg ratio, the flour amount, and the baking temperature. Those five data points tell you the texture, the color, and the chew of the finished cookie before you measure a single ingredient.

• **Stack the variables for maximum impact.** Brown butter plus dark brown sugar plus extra yolks is not magic: it is three separate chemistry decisions all pointing toward the same outcome. You can apply this stacking principle to any cookie style.

• **Troubleshoot cookie problems by variable.** Are cookies spreading too much? Butter was too warm, or there was too much white sugar. Cookies too cakey? Too many whites, too much flour, too much baking powder. Cookies pale and bland? Try an overnight rest, switch to dark brown sugar, or raise your oven temperature. You now have a diagnostic framework, not a guessing game.

9 Cakes, Foams, Batters, and Tenderness

Diagram 9.1: The Cake Kitchen, Three Families on One Counter

Why Cakes Are a Science Problem

A cake is a stabilized foam baked into a permanent structure. That single sentence predicts almost everything about how cakes work and why they fail.

Foam means gas trapped inside a liquid or semi-liquid matrix. In cake batter, the gas comes from three possible sources: air you whip in mechanically during mixing, carbon dioxide released by chemical leaveners, and steam from the water in your ingredients.

The job of the other ingredients is to harden around those gas bubbles before they escape. Eggs coagulate and set a protein network. Flour provides starch that gelatinizes into a firm scaffold. Sugar delays both processes, so the structure does not lock before the gas bubbles have finished expanding. Fat lubricates the system.

Every decision in a cake recipe is a decision about foam. "Cream until pale and fluffy" is a foam-building instruction. "Do not overmix" is a foam-protection instruction. "Bake at 350F" is a foam-to-structure timing instruction. Once you see the foam, you see the whole cake.

The Three Families

Cakes separate into three broad families based on where their air comes from and what role fat plays.

The first family is butter cakes: pound cakes, layer cakes, and most celebration cakes—these start by whipping air into softened butter and sugar. The fat carries the foam. Eggs add water and emulsifying power. Flour and leaveners build the scaffold. Butter cakes are rich and tender, with a characteristic fine crumb from the dense network of fat-coated bubbles laid down during creaming.

The second family is foam cakes: genoise, sponge, angel food, and chiffon. These build their foam from eggs or egg whites, not butter. The mixing method becomes everything. Fold wrong, and you lose the foam. Foam cakes use little or no chemical leavener. Physics does the work.

The third family is oil cakes: carrot cake, zucchini cake, olive oil cake, and many coffee cakes. Oil is liquid at room temperature, so there are no fat crystals to hold bubbles in place. These cakes rely entirely on chemical leaveners for their rise. What oil sacrifices in lift, it makes up for in moisture retention, because liquid fat coats the starch granules without solidifying. Carrot cake stays moist for three or four days. Olive oil cake tastes better on day two.

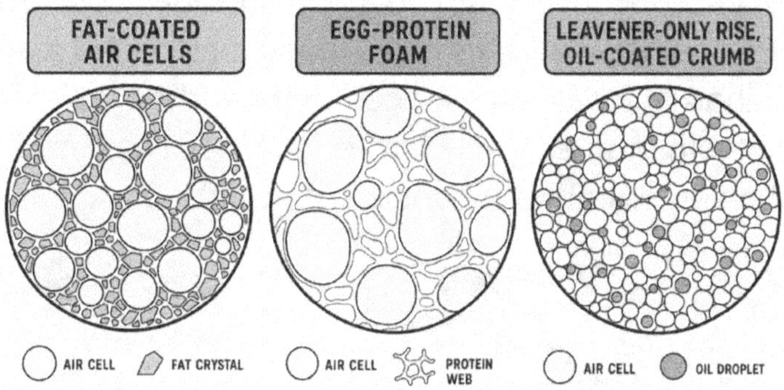

Diagram 9.2: Three Cake Families, Air Sources Compared

The Creaming Method: Building Foam from Fat

Creaming defines butter cakes and is one of the most misunderstood steps in home baking. Most people cream for thirty seconds to two minutes. True creaming takes five to eight minutes in a stand mixer, and the difference is dramatic.

Butter at room temperature, around 65F to 68F, has a crystal structure that is part solid, part liquid fat. When you beat it with granulated sugar, the jagged edges of the sugar crystals cut into the butter, creating tiny pockets. Each pocket fills with air from the mixer paddle. Fat crystals hold each pocket's walls rigid. Beat long enough, and you have millions of fat-coated air cells distributed through the fat matrix.

Those air cells are the backbone of the cake's lift. When the batter enters the oven, the air inside each cell expands as it heats. Carbon dioxide from any baking powder migrates into those preformed cells rather than nucleating new ones. A preformed cell is far more stable than a freshly formed bubble. The result is a fine, even crumb with consistent holes.

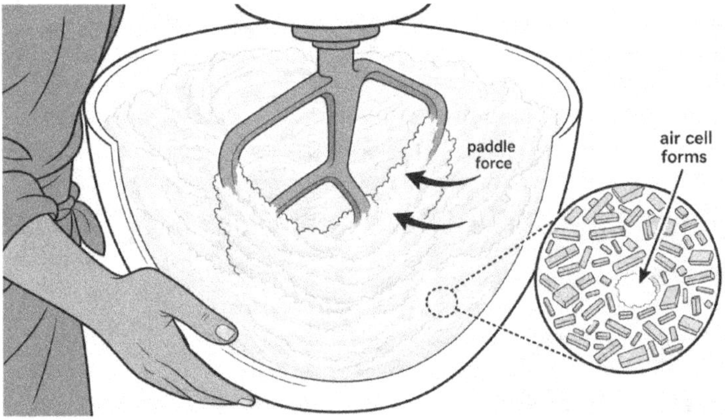

Diagram 9.3: Creaming in Action, Fat Crystals Trapping Air

Butter Cakes: Ratio and Variation

The classic pound cake is the purest expression of the butter cake family. One pound each of butter, sugar, eggs, and flour. The ratio is 1:1:1:1 by weight, and the cake works because of that balance.

Fat tenderizes meat by coating proteins with flour and preventing them from forming long gluten strands. Sugar delays protein setting and starch gelatinization, buying time for the cells to expand. Eggs add structure, lift, moisture, and emulsification through lecithin in the yolk. Flour provides the starch scaffold that sets permanently in the oven. Change any ratio significantly, and the whole system shifts.

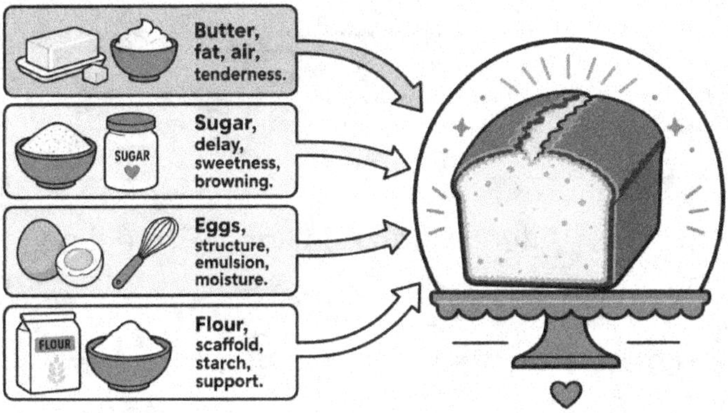

Diagram 9.4: The Pound Cake Ratio, Four Ingredients in Balance

The Alternating Addition Method

Professional layer cake recipes instruct you to alternate dry and liquid ingredients, starting and ending with dry. It looks fussy, but the reason matters.

When you cream butter and sugar and add eggs, you form a fat-continuous emulsion, a system in which tiny water droplets from the eggs are suspended in the fat phase. That emulsion is stable only as long as you do not overwhelm it with too much water at once. Adding all the milk at once pours too much water into the fat phase and breaks the emulsion. The batter curdles and fat pools.

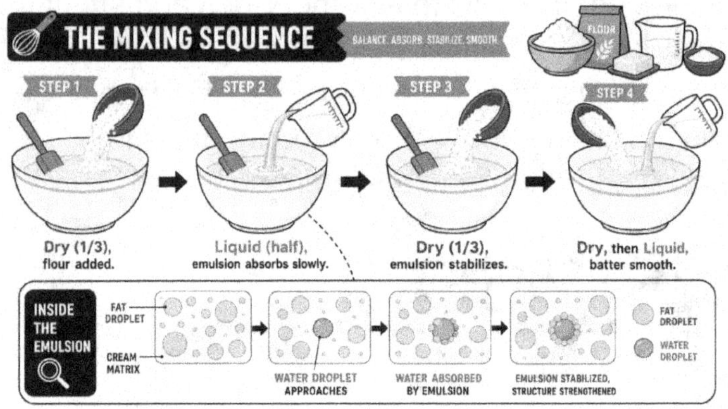

Diagram 9.5: Alternating Addition, Protecting the Emulsion

Egg-Foam Cakes: Physics as Leavener

The foam-cake family depends on eggs for structure and air in almost equal measure. Chapter 5 covered egg foams in detail; this section builds on that foundation.

When you whip whole eggs with sugar, as in a genoise or classic sponge, the proteins partially unfold, stretch into thin films, and trap air. Sugar stabilizes those films by increasing the viscosity of the water phase. The foam is lighter and more fragile than a butter cake batter.

When you whip only egg whites, as in angel food cake, there is no fat from the yolk to interfere. That is why angel food achieves a cotton-candy lightness that butter cake never can. But purity makes the foam fragile. A single drop of yolk sabotages it, because even a trace of fat prevents the proteins from linking—clean bowls, clean beaters, no yolk contamination.

Diagram 9.6: Angel Food Foam, Whipped Whites in a Tube Pan

The Chiffon Middle Ground

Chiffon cake sits at the intersection of the foam-cake and oil-cake families. It tastes rich because oil adds fat, but it feels as light as a foam cake because whipped whites provide the lift.

Combine oil, egg yolks, liquid, and flour into a smooth batter. Separately, whip egg whites with sugar to a medium-stiff peak. Fold the whites into the batter in several light additions. The whites carry the air; the oil carries the moisture; the flour and eggs carry the structure.

Diagram 9.7: Chiffon Method, Whites Folded into Oil Batter

Oil Cakes: Chemical Lift, Long Moisture

Oil cakes are the most forgiving family. There is no creaming phase, no foam building, no dramatic fold. Combine dry ingredients in one bowl, wet ingredients in another, and bring them together with minimal stirring. The goal is to hydrate the flour without developing gluten. Gluten strands form easily the moment wet and dry combine, which is why you stop as soon as the flour disappears.

Diagram 9.8: Oil Cake Method, Wet Meets Dry

Cake Flour and the Gluten Question

Cake flour sits at about 7 to 8 percent protein; all-purpose runs 10 to 12 percent; bread flour runs 12 to 14 percent. In a cake batter, where the goal is maximum tenderness, those few percentage points are the difference between a silky crumb and a tough, bready one.

The reason is gluten, the elastic network that forms when two wheat proteins, glutenin and gliadin, hydrate and link during mixing. In bread, gluten is desirable. In cake, it is largely the enemy. Too much gluten makes the crumb tough, creates tunnels (long vertical channels where gas forces its way through a too-firm structure), and makes the cake difficult to cut cleanly.

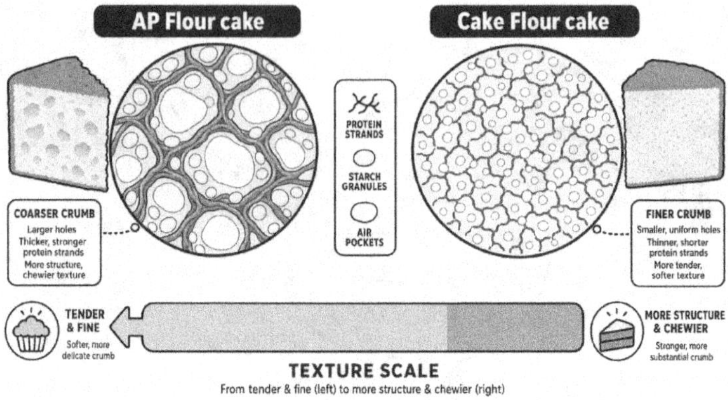

Diagram 9.9: Cake Flour vs. AP Flour, Protein Comparison

Sifting: Why It Matters Here

Sifting is an instruction people often skip. For cookies or muffins, that shortcut rarely causes obvious problems. For foam cakes, angel food, and chiffon, skipping the sift genuinely matters.

Flour from a bag is compacted. A cup of compacted flour weighs measurably more than a cup of sifted flour. If you measure by volume, unsifted flour adds extra weight to the batter: extra gluten-forming protein and extra structure exactly where you do not want it. The result is a denser, sometimes sunken cake.

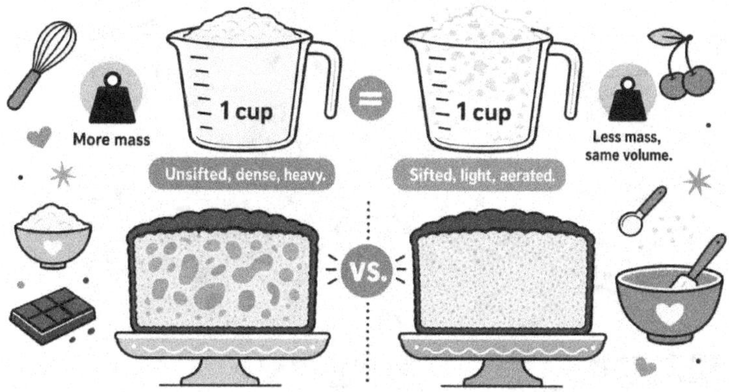

Diagram 9.10: Sifting, Compacted vs. Aerated Flour

The Reverse-Creaming Method

The reverse-creaming method runs the mixing sequence in the opposite order from standard creaming. You start with flour, sugar, baking powder, and salt, add softened butter, and mix until sandy. Then add the eggs and liquid, and mix until smooth.

By coating the flour with butter before adding any liquid, you wrap every grain of flour in a thin layer of fat. That fat layer blocks water from reaching the gluten-forming proteins, resulting in noticeably less gluten development and a more tender, velvety crumb, even when using all-purpose flour.

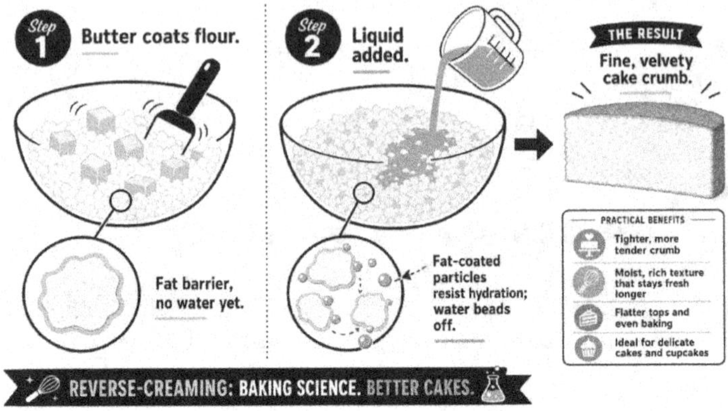

Diagram 9.11: Reverse Creaming, Flour Coated Before Liquid

Overmixing, Tunnels, and Toughness

Overmixing is among the most common and most misunderstood cake failures.

When you add flour to wet batter and stir, gluten forms immediately. The longer you stir, the more the strands align into longer, more elastic chains. Gas bubbles pushing against that elastic network follow the path of least resistance, forcing upward through the batter in a long, narrow channel. That channel is what bakers call a tunnel: a long vertical or diagonal hole in the crumb.

Diagram 9.12: Overmixed vs. Properly Mixed, Crumb Comparison

Leavening Chemistry in Cakes

Even butter cakes that depend heavily on creamed air benefit from chemical leavening. Most layer cake recipes include baking powder, and some include baking soda alongside an acid such as buttermilk.

Baking powder is double-acting: one reaction occurs at room temperature when the powder contacts liquid, and a second, larger reaction occurs when the batter reaches about 140°F in the oven. That second reaction contributes more to the rise, which is why a batter can sit in the pan for a few minutes without losing significant lift.

Baking soda requires an external acid: buttermilk, yogurt, lemon juice, vinegar, brown sugar, honey, or cocoa. It reacts immediately and strongly, so batters with baking soda should go into the oven without delay.

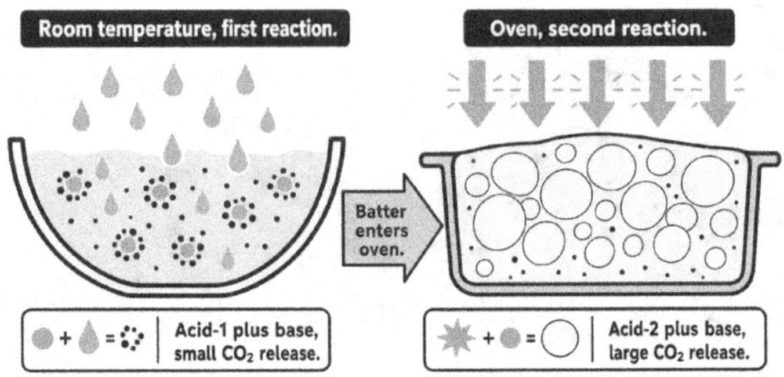

Diagram 9.13: Double-Acting Baking Powder, Two-Stage Release

Doneness and the Interior Structure

Knowing when a cake is done short-circuits anxiety. Science gives you three reliable signals.

The first is the toothpick test. A clean toothpick means starch has gelatinized and eggs have coagulated enough to set the crumb. A wet toothpick means the center is still liquid. Moist crumbs on the toothpick fall in the middle zone: the cake is nearly set and will firm as it cools. That moist-crumb result is often the sweet spot for tender layer cakes.

The second is internal temperature. Starch gelatinization and egg protein coagulation both complete around 180F, but the interior continues to set after you pull the cake from the oven. The target at the very center runs from 200F to 210F. Below 200°F, the center is likely still doughy. Above 210F, you have exceeded the optimal moisture content.

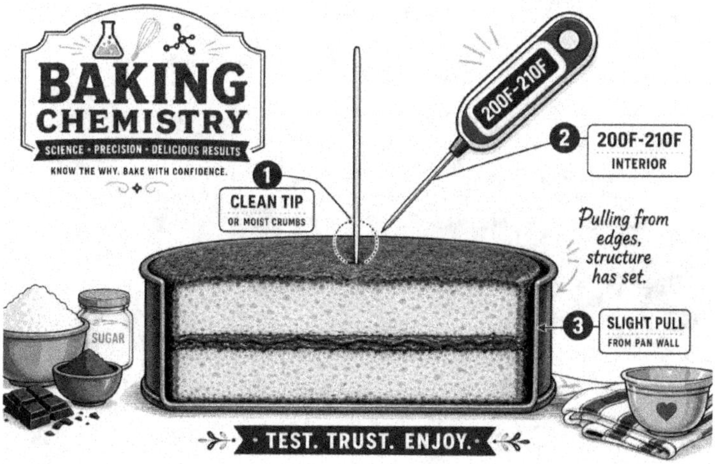

Diagram 9.14: Doneness, Three Signals at Once

Cooling: In Pan vs. On Rack

The structure of a cake continues to set for several minutes after it leaves the oven, so the cooling strategy is not an afterthought.

Most cakes should cool in the pan for ten to fifteen minutes before unmolding. During this time, the internal temperature redistributes, and the starch network completes its setting. A cake unmolded immediately from the oven can compress under its own weight or tear if the crust sticks.

Diagram 9.15: Cooling Steps, In Pan, Then on Rack

Frostings: The Chemistry of the Envelope

A frosting is not a decoration. It is an insulating, moisture-preserving envelope that changes how the entire cake tastes and how long it lasts.

American buttercream is the simplest: powdered sugar beaten into softened butter. The sugar dissolves into the butter, creating a thick, paste-like consistency. Intensely sweet, stable at room temperature, and very forgiving. It can taste grainy if not beaten long enough, and it softens above 72F.

Swiss meringue buttercream starts by whisking egg whites and sugar over a double boiler until the sugar fully dissolves, then whipping to a stiff meringue, cooling it, and beating in softened butter piece by piece. The result is silky and much less sweet. The emulsification of egg white and butter creates a stable, spreadable texture that tastes lighter than its fat content suggests.

Diagram 9.16: Frosting Families, Three Textures Compared

High-Altitude Baking: The Pressure Drop

Bakers above 3,500 feet deal with cakes that rise too fast, expand to twice the expected volume, and then collapse before fully setting.

Air pressure decreases with altitude. Lower pressure means the gases inside your batter face less resistance as they expand. The same carbon dioxide bubble that would grow moderately at sea level expands much further at altitude. The cake over-rises faster than its protein and starch structure can set to hold the expanded shape—the structure tears. The air escapes. The center collapses.

The adjustments follow logically. Reduce the leavening by about one-quarter to produce less gas. Add a small amount of extra flour, one to two tablespoons per cup, to strengthen the scaffold. Increase the oven temperature by 15F to 25F to accelerate starch gelatinization, locking the structure before over-expanded bubbles escape. Reduce sugar slightly, as high sugar levels delay the structure's setting. Add a small amount of extra liquid to compensate for faster evaporation.

None of these adjustments is large on its own, but they work together to push the gas-to-structure timing back in the baker's favor.

Try This in Your Kitchen

The creaming-time test is one of the clearest demonstrations in baking chemistry, and it costs you only one extra cake.

Make two identical butter cake batters from the same ingredients, dividing everything equally for the first; cream butter and sugar for two minutes. For the second, cream for a full eight minutes until the mixture is pale, billowy, and tripled in volume.

Bake both cakes in identical pans at the same temperature and for the same time. The eight-minute cake should be visibly taller with a more even dome. Slice both in half. The two-minute cake will show larger, irregular holes and a coarser texture. The eight-minute cake will show a finer crumb and taste noticeably more tender, because the fat-coated air cells interrupt gluten formation more effectively.

Those six extra minutes added hundreds of thousands of air cells, each contributing to a finer crumb and a lighter texture. You can feel it in every fork.

What You Can Now Do

• Identify which of the three cake families a recipe belongs to by reading the ingredients and method before you start, and predict how it will behave based on where its air comes from.

• Cream butter and sugar properly, for the full five to eight minutes until pale and tripled in volume, building the foam that determines your crumb's fineness and tenderness.

• Choose between cake flour and all-purpose flour intentionally, use the cornstarch substitution when needed, and understand why sifting is not optional for foam cakes and angel food.

• Fold egg-white foam into batter with the light J-stroke technique, keeping each addition small and each pass deliberate, to protect the air that lifts the entire structure.

• Read doneness from three simultaneous signals: clean or moist-crumbed toothpick, 200F to 210F interior temperature, and the thin gap where the cake has pulled from the pan edges.

• Cool foam cakes inverted so the fragile structure sets before bearing its own weight, and transfer all cakes to a wire rack to prevent steam from condensing into a soggy base.

• Match a frosting to the cake and occasion: American buttercream for ease and speed, Swiss or Italian meringue buttercream for silkiness and stability, ganache for intensity and elegance.

10 Pies and Pastry, the Geometry of Butter

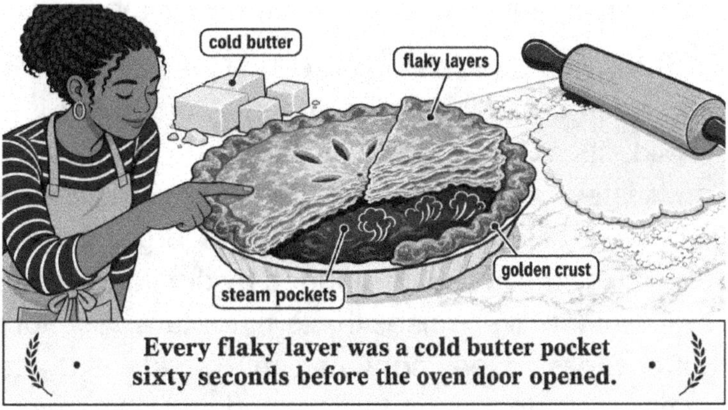

Diagram 10.1: The Flaky Crust Moment, CHAPTER OPENER

The Core Secret: Fat Distribution

Everything in pastry starts with one principle: the size and shape of fat particles inside a dough determines the texture of the finished product.

If fat is present in large, flat chunks, those chunks melt during baking, leaving empty spaces. Steam rushes in, and you get flake. If fat is blended smoothly and evenly into flour, it coats the proteins in the flour and prevents them from bonding into long gluten chains. The result is tender and short, meaning it snaps cleanly, but it has no flake because there are no pockets to open. If fat is melted into water and then cooked into flour before eggs are added, the starch gelatinizes, the eggs set into a network, and the trapped water becomes a steam engine during baking, inflating the pastry into a hollow shell. That is choux.

Three archetype outcomes: flaky, short, hollow-puff. One controlling variable: how the fat meets the flour.

This is the same principle explored in Chapter 4, when we discussed butter's physical states. Cold butter is a rigid solid. Its water molecules (butter is roughly 18 percent water) are locked inside the solid fat crystal matrix. When cold butter sits in flour in large chunks, those chunks stay solid, surrounded by dry flour, intact. The moment the oven raises the temperature past 90F, the butter melts, releasing water, milk solids, and fat. The water turns to steam, the fat lubricates the layers of flour, and the milk solids begin Maillard browning, which is why an all-butter crust tastes more complex and golden than a shortening crust.

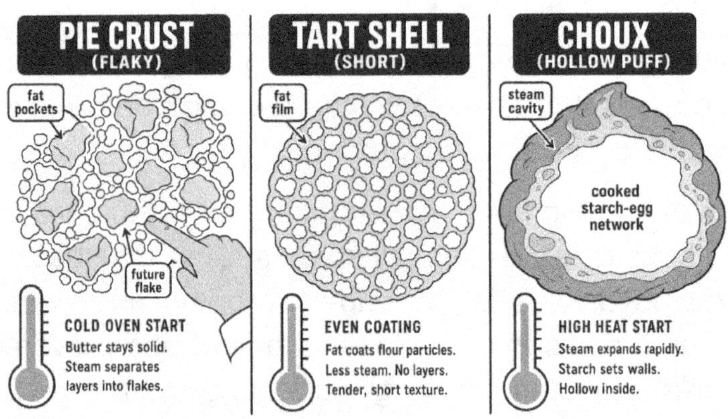

Diagram 10.2: Fat Distribution Compared

Pâte Brisée: Building the Flake

The classic pie crust formula is simple: flour, fat, cold liquid, salt. The method is cutting, not mixing. You work the fat into the flour using a pastry cutter, two knives, or your fingertips, breaking it into smaller, irregular pieces while keeping every piece coated in dry flour. You stop when the mixture looks like a rough crumble, with pieces ranging from pea-sized to oat-flake-sized. That range of sizes is intentional.

Then you add liquid —only just enough, a tablespoon at a time. You flick and toss rather than stir, because stirring encourages gluten development, and gluten is the enemy of flakiness. You want the flour to hydrate barely enough that the dough holds together when squeezed, but no more. Then you press it into a disc, wrap it, and refrigerate it for at least thirty minutes before rolling.

Every step serves the fat geometry. Cutting keeps butter chunks intact. Minimal liquid keeps gluten from forming. Chilling re-solidifies any fat your warm hands softened. Rolling the chilled disc extends the fat chunks into flat horizontal sheets. Those sheets will become your flake.

Diagram 10.3: The Cutting-In Process

Why Cold Is a Non-Negotiable

Cold is not a preference in pie crust. It is chemistry.

At room temperature, around 68F to 72F, butter is soft enough to smear. When soft butter meets flour during cutting-in, it smears, coating proteins the way blended fat coats them in a tart shell. The result is a mealy, dense crust with no distinguishable layers.

When cold butter meets flour, it breaks instead of smearing. Each fragment stays intact, surrounded by dry flour, all the way until the oven. Ice water matters for the same reason. Warm water hits the exposed butter surfaces, softening them, and you lose layers.

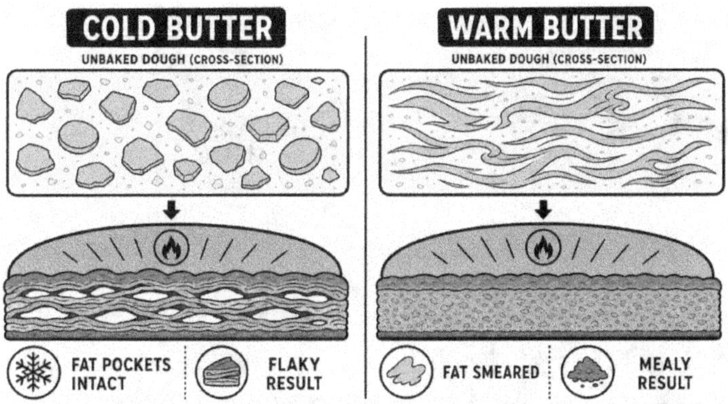

Diagram 10.4: Cold vs. Warm Butter in Dough

All-Butter, Shortening, or Hybrid

Butter is 80 percent fat, 18 percent water, and 2 percent milk solids. The water inflates the flake pockets via steam. The milk solids undergo Maillard browning, giving an all-butter crust its toasty golden color and complex flavor. Shortening is 100 percent fat, with no water and no milk solids. It stays solid at higher temperatures than butter, making it more forgiving in warm kitchens. The flake can be just as dramatic in height, but the layers can feel slightly drier, and the flavor is neutral.

The hybrid approach, typically around three parts butter to one part shortening, tries to capture the best of both. Butter provides flavor, Maillard browning, and internal steam. Shortening slightly raises the dough's melting point, making it more forgiving when rolling.

Diagram 10.5: Butter vs. Shortening Composition

The Vodka and Vinegar Trick

The standard tension in pie crust is this: you need liquid to hydrate the flour enough to hold the dough together, but every extra drop risks activating gluten. Gluten forms when water contacts the proteins in flour (gliadin and glutenin), allowing them to bond into long, elastic chains. Long gluten chains mean chew, which is wonderful in bread and catastrophic in pie crust.

Ethanol, the alcohol in vodka, solves this. Ethanol molecules mix with water and penetrate the flour, but their chemical structure does not bond to gliadin and glutenin proteins the way water does. A tablespoon of vodka in your ice water adds liquid volume, making the dough rollable without triggering the same level of gluten bond formation as an equivalent tablespoon of plain water would. The result is a more forgiving dough that rolls without developing snap-back elasticity.

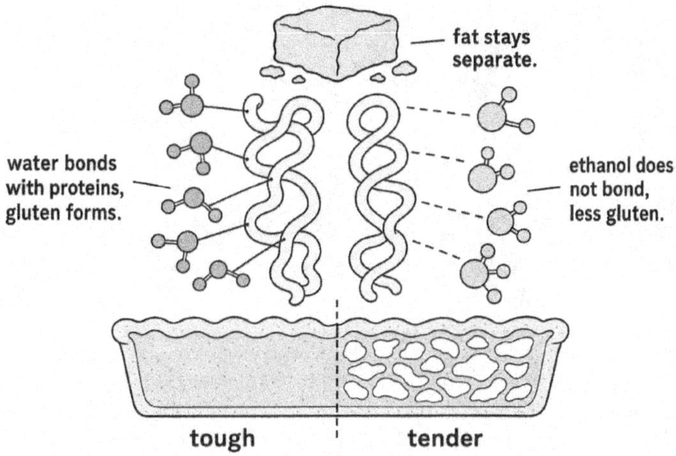

Diagram 10.6: Why Vodka Protects the Crust

The Role of Resting

After mixing, before rolling, the dough rests. This step is framed as "optional" in many recipes, but the chemistry makes clear it is not. Resting does at least four things at once. First, it allows water to migrate more evenly through the flour. Second, it allows gluten strands to relax. Gluten under tension snaps back like a rubber band when you try to roll it. If you have ever tried to roll pie dough that was not rested and watched it shrink back toward the center every time you pressed the rolling pin, you experienced gluten tension firsthand. Thirty minutes of rest largely eliminates this. Third, any fat that warmed during handling re-solidifies into discrete solid chunks. Fourth, the cold temperature is a reset: the dough enters the oven in the best possible fat-geometry condition.

Diagram 10.7: Before and After Resting

Rolling, Layering, and What the Pin Does

Rolling is not just flattening. It is engineering.

When you press a rolling pin into chilled pie dough, you apply pressure to the fat chunks, elongating them horizontally. A roughly round butter chunk becomes an oval, then a flat disc, then a thin horizontal sheet as you roll. Each original butter chunk has become a thin sheet running horizontally through the dough.

Those horizontal sheets are the layers-in-waiting. In the oven, the sheet of solid fat melts. The water embedded in the fat flashes to steam, pushing upward. The now-liquid fat lubricates the flour layers above and below, so they slide apart rather than sticking. The steam pressure inflates the gap. The flour layers set as the starch gelatinizes and the gluten coagulates, locking the inflated shape in place. Each flake you see in a baked pie crust is the ghost of a butter sheet.

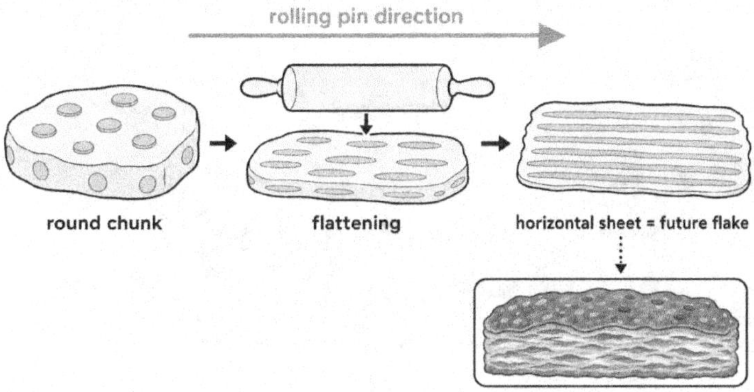

Diagram 10.8: Rolling Pin Geometry

Single Crust, Double Crust, and Lattice

The structural demands on a pie crust change depending on its role.

A single-crust pie, like a cream pie or a lemon curd tart, uses only the bottom shell. The crust needs to hold its shape, support the weight of the filling, and create a firm base that does not absorb the filling liquid. The bottom crust bakes in the most difficult position: flat against the pan, surrounded by filling that insulates it from direct heat, with steam rising from the filling above.

A double-crust pie, like a classic apple or cherry pie, adds a top sheet of dough. The top crust bakes exposed to full oven heat. The classic thumb-pinched or fork-pressed edge creates a mechanical bond that prevents the filling from boiling over during baking.

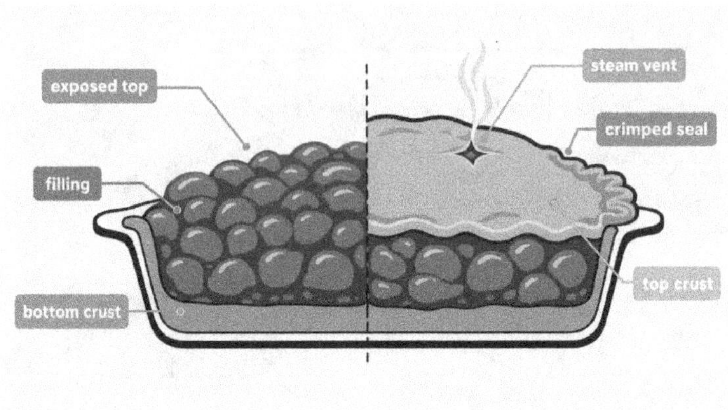

Diagram 10.9: Single vs. Double Crust Anatomy

Blind Baking and the Soggy Bottom Problem

A soggy bottom occurs when the bottom crust absorbs moisture from the filling before it can set and crisp up. A wet filling sitting on top of unbaked dough insulates the crust from below, preventing it from ever reaching the temperature needed to dry out and brown.

Blind baking is the solution for single-crust pies. You bake the crust alone, weighted down, before any filling goes in. The shell goes into the oven empty, lined with parchment paper, and filled with pie weights, dried beans, or uncooked rice. The weights hold the bottom crust flat against the pan so it cannot puff from steam, and they hold the sides of the crust up so they cannot slump inward.

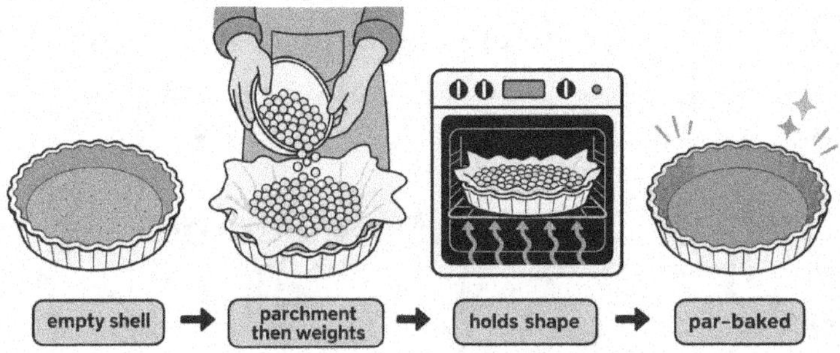

Diagram 10.10: Blind Baking Step by Step

Pate Sucree: The Short, Sweet Shell

Pate sucree (literally "sugared dough") begins with the same ingredients as pie crust but uses a technique borrowed from cookie-making. The butter is softened, not cold. Sugar is creamed into the soft butter. Eggs are added. Flour is stirred in just until combined.

Because the butter was soft when it met the flour, it coated the proteins in the flour before they could form gluten chains. There is essentially no gluten network in pate sucree. The structure of the baked shell comes entirely from egg proteins and gelatinized starch, which is why tart shells snap cleanly with no springiness.

The sugar does double duty. It contributes sweetness, but at the quantity used, it also competes with the proteins in flour for available water. Sugar is hygroscopic, attracting water molecules to itself. The sugar instead captures some of the water that might otherwise have hydrated gliadin and glutenin. Less protein hydration means less gluten. More sugar, shorter texture. This is the same mechanism explored in Chapter 3, which discusses sugar's role in cookie tenderness.

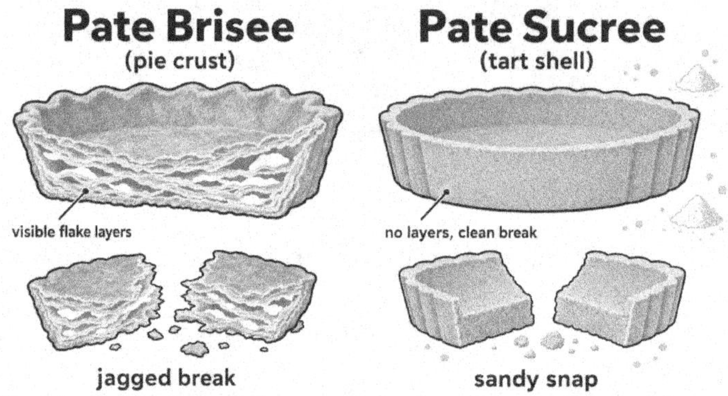

Diagram 10.11: Pate Sucree vs. Pate Brisee Texture

Galettes and Rough Puff

A galette is a free-form pie made without a pan. The dough is rolled out, filled, and the edges are folded up over the filling in a rough, open pleat. Without the pan, the crust holds itself up because folded dough creates multiple layers at the rim that fuse and crisp into a firm, self-supporting edge. The bottom often bakes crisper than the bottom of a pan pie because it has direct contact with the hot metal of the baking sheet.

Galettes are forgiving of dough that is slightly more hydrated or less perfectly cut than a formal pie would tolerate, and many baking teachers recommend them as the entry point for students learning pie dough.

Diagram 10.12: Galette Free-Form Fold

Choux: Steam as Architecture

The third pastry archetype works through a mechanism that sounds almost impossible at first. You boil water and butter together in a saucepan. You add flour all at once and beat the mixture over heat until it forms a smooth, stiff paste and pulls away from the sides of the pan, cooking the starch into a thick, gelatinized matrix. Then you remove the pan from the heat and beat in eggs one at a time. The paste looks slippery and wet but absorbs each egg, becoming smooth and pipeable.

The water content of the paste turns to steam almost immediately. The gelatinized starch and the partially set egg network create a tight, extensible shell. The steam inflates that shell to two or three times the piped size. As it expands, the egg proteins cook and lock the shell in its expanded shape, leaving a large, mostly empty cavity ready to receive cream, custard, or ice cream.

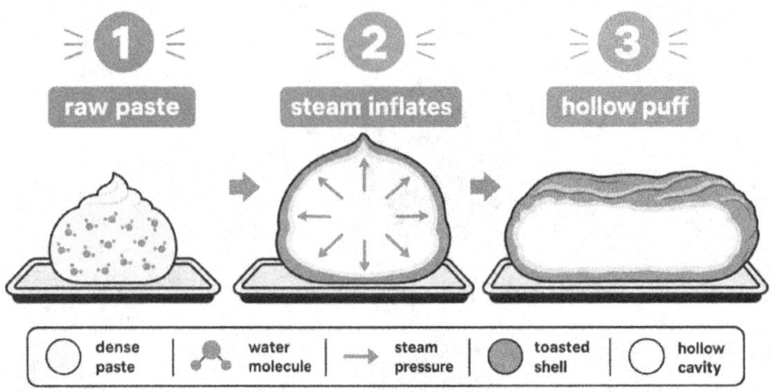

Diagram 10.13: Choux Steam Inflation

The Pre-Bake Temperature Game

Each pastry type has its own oven temperature logic.

Pie crust wants moderate heat, 375F to 400F, sustained long enough to bake through the thick bottom without burning the edges. Edges always brown faster because they are thinner and more exposed. A pie shield slows edge browning while the center catches up.

Tart shells bake at 350F to 375F, but pâte sucrée has more sugar, so the risk of overbrowning is higher. Sugar caramelizes at around 320F and burns at around 375F. A tart shell can go from perfectly golden to dark and bitter in two minutes. Many pastry chefs bake them at 325F and extend the bake time to compensate.

Choux starts hot. Most choux recipes begin at 425°F to 450°F for the first 10 minutes. This high initial heat turns the surface water to steam almost immediately, creating the pressure needed for rapid expansion before the shell can set. If the initial heat is too low, the shell sets before it has inflated fully, and you get a small, dense puff with a thick wall instead of a large, hollow one. After the initial hot blast, the temperature drops to around 375F to finish baking and dry the interior.

Diagram 10.14: Oven Temperature Profiles for Three Pastries

Why Dough Shrinks During Baking

If you have ever pressed a pie crust into a pan, crimped it beautifully, and then watched it shrink and slump during baking, you know this pain. It is the most common complaint from home pie bakers after soggy bottoms.

Shrinkage happens because of gluten. The gluten networks formed during mixing and rolling are under tension. During baking, as the dough heats and the fat melts, it becomes soft enough for the gluten to contract.

The crust pulls away from the pan edges, and the walls slump inward.

The solutions are cumulative. First, rest the dough properly before rolling. Second, do not stretch the dough to fit the pan. Drape it loosely and let it settle under its own weight. Any dough that was stretched is under tension and will snap back. Third, chill the lined pie pan for 15 minutes after fitting the dough, before baking, giving the dough's gluten one more chance to relax. Fourth, use pie weights during blind baking to physically hold the walls up while the crust sets. Most home bakers skip one or two of these steps, usually the second rest after fitting the pan. It is not the recipe. It is the process.

Try This in Your Kitchen

This experiment takes about two hours total. You will make two small test crusts from the same batch of dough.

Divide a batch of pie dough into two equal pieces. Roll the first piece right away, without resting. Fit it into a small tart pan or shape it as a free-form galette round. Notice how the dough springs back when you roll it.

Wrap the second piece and refrigerate it for forty-five minutes. Then roll and fit it the same way. Notice how much more willingly it stretches.

Blind-bake both pieces side by side at 375°F, in the same oven, on the same shelf. After baking, compare the edges. The un-rested crust will have shrunk noticeably more. The rested crust will hold closer to its original size. You have just photographed gluten tension and gluten relaxation in baked form.

Diagram 10.15: The Two-Crust Experiment

What You Can Now Do

• You can explain why pie crust is flaky and tart shell is sandy: fat chunk size and blending method determine whether you get separate layers or uniform tenderness.

• You can choose the right fat for your goal: all-butter for maximum flavor and Maillard color, shortening for a stable high-flake structure in warm kitchens, or a hybrid for the practical middle ground.

• You can troubleshoot a tough crust by asking two questions: was the water too warm, and did you handle the dough too long? Both build gluten, and both are fixable next time.

• You can use vodka or vinegar in your pie water deliberately, knowing that ethanol reduces gluten bonding and acid disrupts protein linking, each giving you a wider margin of error without compromising the finished crust.

• You can blind bake a crust properly, using two bake phases and egg-white brushing to defeat the soggy bottom without guesswork.

• You can explain what choux pastry is doing at the molecular level: cooked starch and egg proteins form an extensible shell, steam pressure inflates it, and the shell sets in its expanded shape, leaving a hollow interior ready to fill.

• You can build shrink-resistant pie crusts by resting the dough twice (after mixing and after fitting the pan), draping rather than stretching, and using weights during blind baking, giving gluten every opportunity to relax before the heat locks the shape permanently.

11 Breads and Laminated Doughs, the Long Game

You measure flour into a bowl. You add water, a pinch of salt, and a tiny scrape of yeast. You stir until no dry flour remains, cover the bowl, and wait. Hours later, the dough has tripled. The surface is alive with bubbles. When you lift the edge with wet fingers, the dough pulls into long, translucent ribbons that stretch, hold, and then slowly release. You have not kneaded anything. You have not turned on the heat. Time did all of this.

No other baking category asks this much patience. Cookies take thirty minutes. Cakes take an hour. Bread takes a minimum of six hours, and a proper sourdough might sit in your refrigerator for eighteen hours before it sees the oven. That wait is precision. Every hour of fermentation produces flavor compounds that cannot be rushed into existence. The flour is being digested. Acids are building. The gluten network is organizing itself strand by strand.

This chapter covers lean doughs (flour, water, yeast, salt, and nothing else), enriched doughs (which add fat, eggs, milk, and sugar), sourdough (wild microorganisms replacing commercial yeast), and lamination (a block of cold butter folded into dough dozens of times to create croissants and puff pastry). By the end, you will understand not just how to bake a loaf, but why everything that happens to it happens.

What Yeast Actually Does

Yeast is a single-celled fungus whose only job in your dough is to eat and breathe. It feeds on simple sugars and exhales carbon dioxide gas. That gas inflates your bread.

Where does the sugar come from in a plain dough with no added sweetener? Raw flour contains starch, which is a long chain of glucose molecules locked together. Flour also contains naturally occurring enzymes called amylases. Amylases are molecular scissors that snip starch chains into shorter pieces, eventually producing maltose and glucose, both of which yeast can eat. This is why bread dough does not need added sugar. The flour provides its own fuel, slowly, over the first hour or two of fermentation.

When you add sugar directly (as enriched dough recipes often do), you give yeast a faster head start. The yeast can eat that glucose immediately without waiting for amylases to work. This is one reason enriched doughs often rise faster than lean ones.

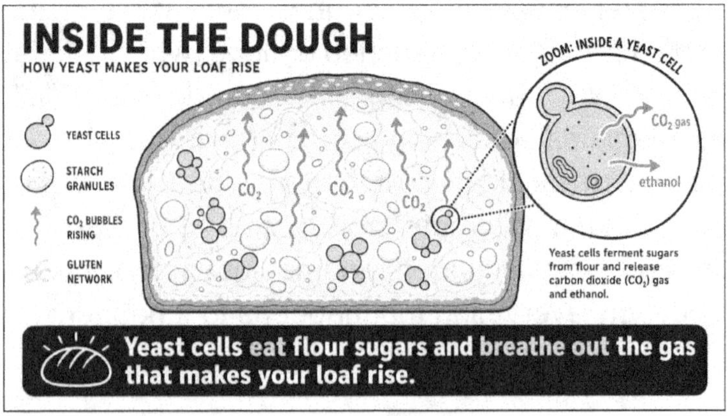

Diagram 11.1: Yeast at Work Inside the Dough

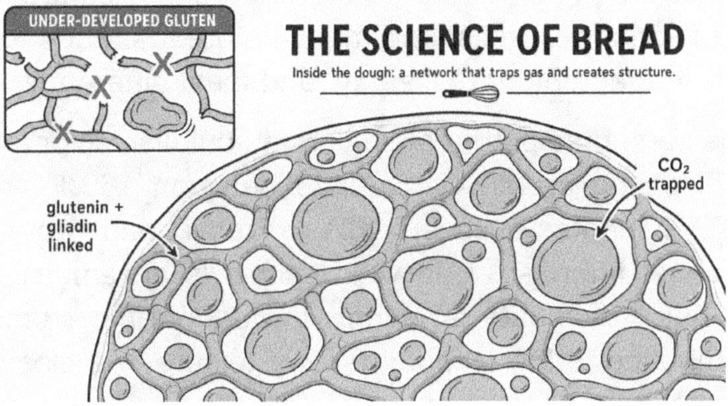

Diagram 11.2: Gluten as a Gas Trap

Lean Dough vs. Enriched Dough

The four-ingredient lean dough is the foundation of artisan bread, baguettes, ciabatta, and most sourdoughs. Flour, water, yeast, and salt- nothing else. Because there is no fat to interfere with gluten development, the gluten forms freely. Because there is no added sugar to accelerate the yeast, fermentation happens at a natural pace. The final texture is chewy, open, and relatively firm, with a crust that crisps hard and stays crisp for hours.

Add fat, and you change everything. Fat coats the protein strands in flour and physically interrupts gluten bonds from forming. This softens the crumb and also means you need longer mixing or more time to develop adequate structure despite the interference. Fat also shortens the shelf life of crust crispness: a bread baked with butter will lose its crunch faster than a plain, lean loaf, because fat attracts moisture from the air. Enriched doughs include brioche, challah, sandwich bread, cinnamon rolls, and many flatbreads. They feel more tender, cut more cleanly, and toast differently.

Eggs contribute both protein and fat, but also color and lift. The egg proteins coagulate between 160°F and 180°F (Chapter 5 covers this in detail) and provide structural firmness. Egg yolk lecithin is an emulsifier that helps distribute fat evenly throughout the crumb, contributing to the smooth, pillowy texture of a classic brioche.

Diagram 11.3: Lean vs. Enriched Dough Side by Side

The Four Phases of Bread

Every loaf, lean or enriched, follows the same four-phase arc: mix, bulk ferment, shape, and proof. Understanding each as a distinct chemistry event helps you diagnose what went wrong when a loaf underperforms.

Mixing hydrates the flour and develops gluten through mechanical action. For lean doughs, you can choose intensive kneading (ten to twelve minutes by machine) or a no-knead approach where time and gentle folds build the structure. Both arrive at the same place: a smooth, elastic dough that passes the windowpane test.

The windowpane test verifies gluten development without any equipment. Pull a small piece of dough and stretch it gently between your fingers. Well-developed gluten stretches into a thin, nearly transparent sheet without tearing. You can see light through it. If it tears, the gluten is still weak.

Diagram 11.4: The Windowpane Test

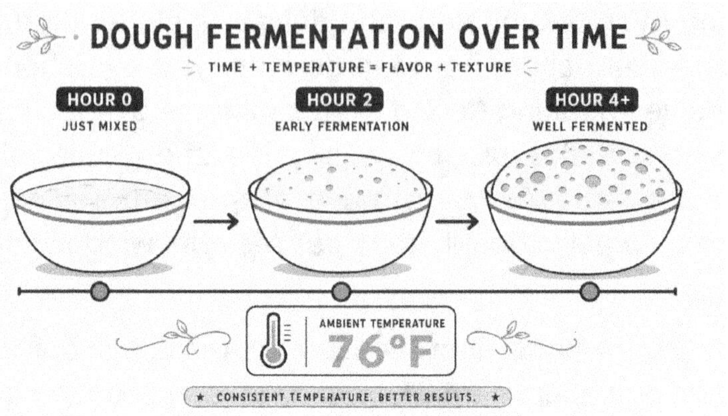

Diagram 11.5: Bulk Ferment Timeline

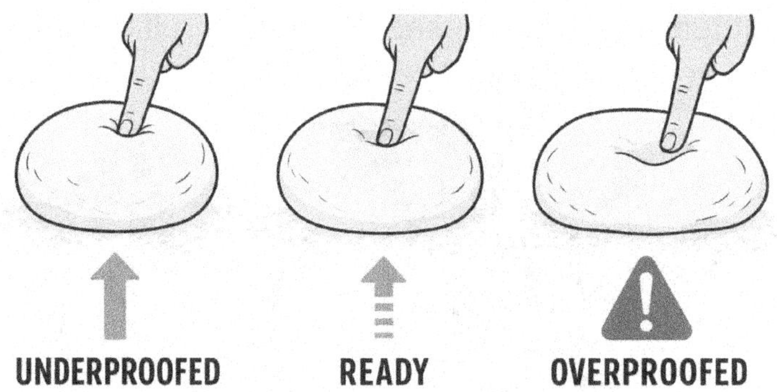

Diagram 11.6: The Poke Test, Three Outcomes

Scoring and the Science of the Ear

Before a proofed loaf goes into the oven, you take a razor blade or a lame and slash the top in one decisive motion. Scoring is a pressure release valve. During the first ten to fifteen minutes in a hot oven, yeast gets one final burst of activity as the temperature climbs from ambient toward the yeast death point at 140F. In this burst, called oven spring, the loaf expands rapidly. Gas bubbles expand with heat. Water turns to steam and pushes outward. All this expansion needs somewhere to go. Without a score, the pressure tears the crust at random. A score directs all that energy deliberately, producing the dramatic "ear" you see on artisan loaves.

The angle of the blade matters. A blade held nearly vertical produces a wide, flat cut. A blade held at 30 to 45 degrees cuts under the skin, and that undercut flap peels back and curls into a raised ear as the loaf springs in the oven. A single decisive slash is always better than a tentative, shallow one.

Diagram 11.7: Scoring Angle and the Ear Formation

Crust Chemistry: Color, Shine, and Crackle

The color and texture of a bread crust depend on the oven temperature, the dough's sugar content, and the wash applied before baking. Maillard browning occurs when amino acids react with reducing sugars above about 280°F, producing hundreds of aromatic flavor and color compounds. Caramelization, the browning of sugars alone, begins above about 320F and contributes its own sweet, rich aromas. Both reactions accelerate as the surface dries out, which is why dry oven heat (after the steam phase) is needed to push the crust to full color.

The wash you apply before baking dramatically shapes the finish.

An egg wash brings protein and fat to the surface. Both participate in Maillard browning and together produce a deep, glossy, lacquer-like finish. Egg yolk alone gives the deepest, richest color. Egg white alone gives gloss without the deep mahogany.

A milk wash is gentler. Milk contains lactose and protein, which contribute to browning, but its lower fat content compared to egg yolk produces a matte, soft finish rather than a high-gloss lacquer. Dinner rolls and soft sandwich loaves use a milk wash for golden, inviting color without high shine.

A water wash does something entirely different. Water adds no color or browning on its own. What it does is stay liquid on the surface during the early baking phase, keeping the crust extensible and allowing it to crack and split beautifully as the loaf expands. A water mist creates the shattering, rustic, crackled crust of a baguette.

Diagram 11.8: Three Washes, Three Crust Outcomes

Sourdough: Wildness and Patience

Commercial bread yeast is a single, selected strain of Saccharomyces cerevisiae, chosen because it is fast, predictable, and produces a clean flavor. Sourdough uses a starter instead: a living culture of wild yeasts and lactic acid bacteria (LAB) maintained by feeding it flour and water on a schedule.

The wild yeasts in a sourdough starter are slower and less aggressive than commercial yeast. The LAB that live alongside them produce lactic acid and acetic acid as byproducts of fermentation. These acids make sourdough taste sour. Lactic acid produces a mild, yogurt-like tang. Acetic acid is sharper and more vinegary. The balance depends on temperature, hydration, and feeding frequency. A cold, stiff starter produces more acetic acid. A warm, wet starter produces more lactic acid.

Book 3 will cover the complete fermentation chemistry in depth. The key insight here is that sourdough is a community, not an individual. The flavors in a sourdough loaf are the product of multiple microorganisms working at different speeds over eight to twenty-four hours.

Diagram 11.9: Inside a Sourdough Starter

Diagram 11.10: Sourdough Fermentation Arc

Mixing, Kneading, and the No-Knead Method

For much of baking history, developing gluten meant kneading: pushing dough, folding it, turning it, pushing again, for fifteen to twenty minutes by hand. This mechanical action pulled protein strands apart, allowing them to realign into a stronger network. It still works. But it is not the only path.

The no-knead method works on a different principle: time and hydration. If you mix a very wet dough (70 to 80 percent hydration) and leave it at room temperature for twelve to eighteen hours, the water does the mechanical work. Proteins hydrate slowly, glutenin and gliadin link without being forced, and the resulting dough develops strong gluten with no kneading at all. The long fermentation also develops flavors that kneaded doughs often lack. The tradeoff is that no-knead doughs require high hydration, making them sticky, and long fermentation times, making them impractical for same-day baking.

Diagram 11.11: Stretch-and-Fold Sequence

The Enriched Dough Universe

Think of the lean-to-enriched spectrum as a dial. At one end: a baguette at 62 to 65 percent hydration, no fat, no sugar, no eggs. Turn the dial slightly, and you get focaccia, still lean but with olive oil creating a richer crumb and crisper bottom. Keep turning, and you arrive at challah (eggs and oil), then brioche (eggs and a very high butter ratio, often 50 to 60 percent butter by weight of flour).

Brioche is the most enriched bread commonly used. Because of its extremely high butter content, the fat so thoroughly inhibits gluten formation that the only thing holding it together is primarily egg protein. The result is a crumb that is extraordinarily tender, almost cake-like, with a deep gold color from egg yolks and a warm butter smell that is hard to resist.

Challah uses eggs but typically no butter. The eggs contribute color and lift, the oil contributes fat and tenderness, and the braided shape creates a beautiful layered surface that takes an egg wash to a deep, lacquered mahogany. The braid is not just decorative: the pull-apart layers allow portions of the loaf to be torn away naturally, without cutting.

Diagram 11.12: The Enrichment Spectrum

Lamination: Layers Made by Geometry

Laminated doughs live in an entirely different world from yeasted bread. The leavening is not fermentation. It is steam. And the technique used to create the staggering layers of pastry is purely physical: you repeatedly fold cold butter into the dough, creating a geometric multiplication of alternating butter-and-dough layers.

A single letter fold takes three layers and turns them into nine. Three of these folds, applied sequentially with resting time between each, turn nine into twenty-seven, then eighty-one, then two hundred forty-three. A classic croissant typically undergoes three letter folds, resulting in roughly 81 layers of butter and dough. Classic puff pastry often undergoes six folds, creating 729 or more layers. The geometry of exponential multiplication is always the same.

The butter in a laminated dough is not mixed in. It is encased inside the dough. The process begins by creating a butter block: softened but still cold butter beaten into a flat, even slab, then chilled again until firm but pliable. This block is placed in the center of a rolled-out dough sheet, and the dough is folded over to encase the butter, like an envelope completely. This package is called the lock-in. From here, every fold multiplies the layers.

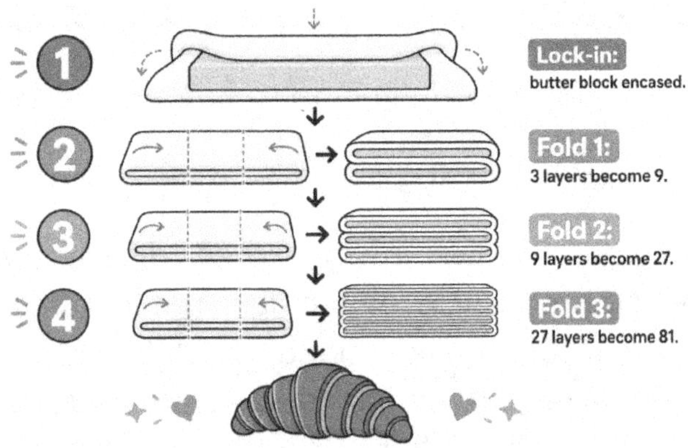

Diagram 11.13: Lamination Geometry, Three Folds

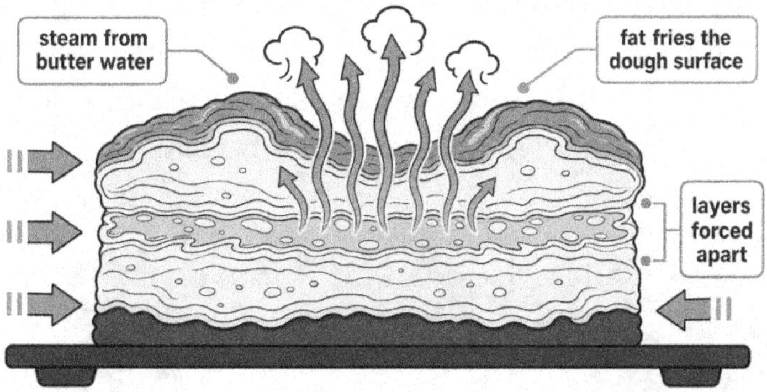

Diagram 11.14: Steam Blowing Layers Apart in the Oven

Temperature Is Everything in Lamination

Home bakers fail at lamination for one reason more than any other: the butter gets too warm.

The butter in a laminated dough must stay cold enough to remain solid throughout all the rolling and folding. If it warms and softens, it no longer stays in discrete layers. Instead, it smears into the dough and gets absorbed. You lose the layer structure. The croissants that come out of the oven look like bread rolls, not shatteringly flaky pastry. The telltale sign is greasiness: dough that feels heavy and slick rather than light and stiff.

The target temperature range for lamination is 50F to 60F for both dough and butter. At 50F, butter is firm but slightly pliable, like modeling clay rather than a rock. If butter is too cold (below about 45F), it will shatter and crack rather than bend when you fold it. Cracked butter tears through the dough, breaking the layered structure. If butter is too warm (above about 65F), it smears immediately. The window is narrow.

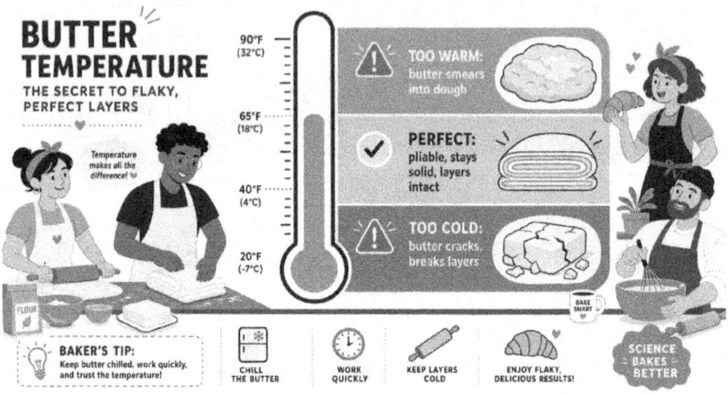

Diagram 11.15: The Butter Temperature Window

Try This in Your Kitchen

Make two simple focaccia doughs using identical ingredients: flour, water, yeast, salt, and a generous pour of olive oil. Mix both at the same time with the same technique. Place one in a warm spot (75F to 80F) and let it ferment for four hours at room temperature. Place the second in your refrigerator for twelve to eighteen hours overnight.

The next day, bake both at the same temperature in the same oven for the same amount of time. Let them cool completely before tasting. Taste them side by side, slowly, paying attention to the crumb structure (bubble size and regularity), the crust texture, and most importantly, the flavor: mild and yeasty, or complex, slightly tangy, and deeply savory?

Almost every taster notices that the overnight focaccia tastes more interesting. The flavor is more layered, with a background acidity that adds depth. The crumb often shows larger, more irregular bubbles because slow fermentation gives the gluten more time to relax and the gas more time to organize into fewer, larger pockets.

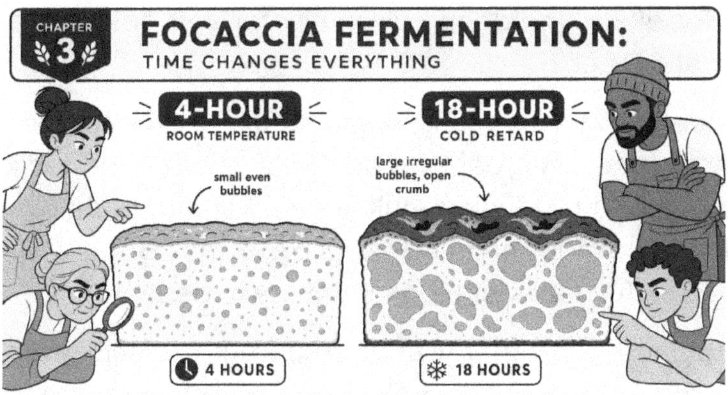

Diagram 11.16: Short Ferment vs. Long Ferment Focaccia

What you can now do

• Recognize the difference between lean and enriched doughs by their ingredient list and predict how each will behave in terms of texture, color, crust, and shelf life.

• Describe what yeast does during fermentation in plain terms: it eats sugars, produces CO2 that inflates gluten-trapped bubbles, and generates flavor compounds that develop over many hours.

• Use the windowpane test to confirm gluten development and the poke test to judge proofing readiness, so you can bake by feel rather than by timer alone.

• Explain why scoring a loaf before baking directs oven spring into an intentional ear rather than a random split, and why blade angle determines whether you get a flat cut or a raised flap.

• Choose the right crust wash for your desired finish: egg for gloss and deep color, milk for soft gold, water for rustic crackle.

• Understand why sourdough tastes more complex than commercial yeast bread (wild yeasts plus lactic acid bacteria producing multiple acid compounds over a slow, long fermentation) and know that Book 3 will cover the full fermentation chemistry in depth.

• Explain the geometry and temperature discipline of lamination: butter stays solid in discrete layers, steam blows those layers apart in the oven, and the critical factor is keeping both dough and butter in the 50F to 60F window throughout every fold.

12 Baking Myths Decoded

You are standing at the kitchen counter, flour on your hands, reading two different blog posts on your phone. One says, "Never open the oven, or your cake will fall." The other says, "If you don't sift your flour three times, your crumb will be dense." Your mother-in-law calls from the living room: "Keep it quiet in there, the cake can hear you." You look at the oven. You look at the flour bag. And you wonder: what is actually true?

That question is what this chapter is for. By the time you reached Chapter 12, you had worked through gluten networks, leavening chemistry, sugar behavior, fat functions, egg architecture, dairy science, oven physics, and bread fermentation. You built a real toolkit. Now you get to use it. This chapter takes the most persistent baking myths and runs each one through the chemistry you already know. Some myths are completely true. Some are completely false. Most land in the middle, where the answer depends on what you are baking, which stage of the bake you are in, and how much of a particular variable you apply. The goal is to understand why a thing might have been true once, why it still matters in some situations, and when you can safely skip it.

Diagram 12.1: Chapter Opener: The Myth Kitchen

Myth 1: Never Open the Oven Door

Opening the oven door is treated like breaking a sacred seal. The cake will fall. The soufflé will collapse. Some bakers press their faces against the oven glass rather than crack the door for half a second. The myth has a grain of truth, but it gets the timing completely wrong for most baked goods.

A standard home oven at 350F can drop by 50F or more in the first 10 seconds after the door is opened. In the first third of baking for most cakes and quick breads, the batter is still liquid, and the leavening gases are still expanding. A sudden temperature drop at this stage can slow that expansion before the gluten-and-egg-protein matrix sets. The batter sags back, resulting in a dense, sunken interior. For a soufflé, which is a foam of egg whites set by heat, interrupting the heat at the wrong moment absolutely matters.

But most of the time, when you actually want to open the oven, your bake is already past that window. You are opening the door at minute 35 of a 45-minute cake bake. By that point, the proteins have set, the starches have gelatinized, and the structure is locked. A brief door opening causes a momentary temperature wobble that recovers in two to three minutes.

The calibrated truth: do not open the oven in the first 20 minutes of a cake, quick bread, or soufflé bake. After that, brief openings are fine. For cookies and pastry, the restriction barely applies.

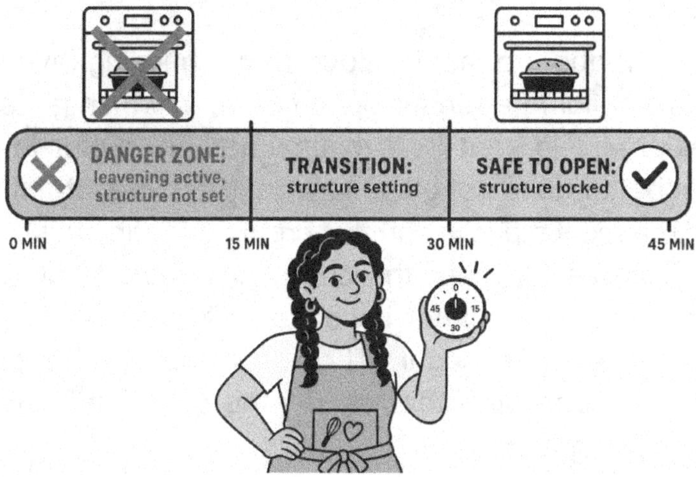

Diagram 12.2: When It's Safe to Open the Oven

Myth 2: Cold Butter Makes Every Crust Flakier

"Keep your butter cold." "Work fast before it warms up." Walk into any baking supply store, and you'll find this advice everywhere. If you are making a flaky pie crust, it is excellent. If you are making muffins or a butter cake, it works against you.

The science behind cold butter in pie crust comes from layering. When you cut cold butter into flour, the butter stays in distinct pieces rather than fully coating the flour. Each piece creates a physical barrier between layers of dough. In the oven, the water in the butter turns to steam, puffing the layers apart and leaving discrete, flaky sheets. This is why pie crust and laminated doughs depend on cold butter: the fat pieces must remain intact until heat converts their moisture to steam.

Now think about a muffin. You are not building layers. You are building a tender open crumb in which gluten development is kept to a minimum. Cold butter in a muffin batter does not blend smoothly into the fat-flour-liquid matrix. It stays in chunks, creating tough pockets. Cake recipes that use the creaming method call for room-temperature butter because soft butter traps air when beaten with sugar. Those air bubbles are the leavening head start that makes the cake light. Cold butter does not cream.

The calibrated truth: cold butter is a specific tool for a specific job. It belongs in pie crust and laminated pastry. It does not belong in cakes, muffins, or quick breads where even fat distribution matters most. Knowing which category your bake falls into is all you need.

Diagram 12.3: Cold Butter vs. Soft Butter by Bake Type

Myth 3: Always Sift Your Dry Ingredients

This myth endured because it was genuinely necessary before modern flour milling. Before the early twentieth century, flour was sifted to remove debris and aerate it after it had been compressed during storage. For those bakers, sifting was practical quality control, not superstition.

Today's commercially milled all-purpose flour is pre-sifted at the mill and arrives at your kitchen already fluffier than a home sift would produce. Spooning your flour into the measuring cup rather than scooping it and whisking the dry ingredients together achieves most of the benefits of sifting without the extra step.

Two cases where sifting still earns its keep: cake flour, which has a much finer particle size and clumps aggressively in storage, genuinely benefits from sifting before it goes into a genoise or angel food batter. And powdered sugar, cocoa, and matcha compact into hard lumps that do not dissolve uniformly; sifting breaks them before they become dry specks in your finished product. For yeast breads, sifting is irrelevant: vigorous kneading breaks up any flour clumps, and gluten development in bread-making works against the light-handed aeration that sifting provides.

The calibrated truth: sift cake flour and powdered-sugar-heavy mixtures. Whisk your all-purpose flour and skip the sifter. Do not sift into yeast doughs.

Diagram 12.4: When Sifting Matters vs. When It Doesn't

Myth 4: Soda and Powder Are Not Interchangeable

You are halfway through a recipe, and the baking powder tin is empty. Baking soda is right there. They are both white powders that make things rise. Surely you can swap them? This swap fails in the mixing bowl and then loudly in the oven, and the reason is acid-base chemistry.

Baking soda is pure sodium bicarbonate. On its own, it is inert; it requires an acid to release carbon dioxide. When you combine it with buttermilk, yogurt, lemon juice, or molasses, the acid in those ingredients reacts, producing bubbles. Baking powder is baking soda with a built-in acid already mixed in, plus a starch to keep them dry and separate until moisture activates the pair. Most baking powder sold today is double-acting, releasing a small burst when wet and a larger burst at oven heat.

When you substitute baking soda for baking powder without adjusting anything else, you have sodium bicarbonate with no acid partner. Very little gas is produced, the baked good is flat, and an alkaline residue leaves a faintly soapy taste. The reverse swap also fails: you need roughly three times as much baking powder to match the leavening strength of a smaller amount of baking soda, and the extra acid salts affect flavor and browning.

The calibrated truth: these leaveners are not interchangeable without adjusting the recipe's acid content. A straight one-for-one swap in either direction will usually not work.

Diagram 12.5: Baking Soda Needs an Acid Partner

Myth 5: Yeast Needs Sugar to Activate

Pour active dry yeast into warm water, toss in a spoonful of sugar, and watch the foam begin to form. That foam feels like proof: the sugar woke the yeast. This is one of the most visually persuasive myths in baking because the evidence appears right in the bowl. Nearly every beginner bread class teaches it. The only problem is that the yeast would have been just as happy without the sugar.

Yeast is a single-celled fungus that ferments carbohydrates and produces carbon dioxide and ethanol as byproducts. Flour is full of starch, a long chain of glucose that amylase enzymes in the flour break down into simpler sugars the moment water is added. The yeast feeds on those sugars just as happily as it feeds on sugar you added directly to the water. What the sugar accomplishes in the warm-water step is to speed up visual confirmation: you see foam sooner. That makes it useful as a check for whether your yeast is alive, not as a requirement for it to function.

Temperature matters far more than sugar. Yeast becomes sluggish below 60F, active around 70F to 80F, and peaks near 95F to 100F. Above 120F, the proteins in yeast cells denature, and the yeast dies. If your water is too hot, no amount of sugar will save the yeast.

The calibrated truth: yeast does not require added sugar to activate. Flour provides all the carbohydrates yeast needs. A small amount of sugar in the proofing water provides a quick confidence check, nothing more.

Diagram 12.6: Yeast Feeds on Flour Sugars Too

Myth 6: Room-Temperature Eggs Always

Ask any baking teacher whether eggs should be at room temperature, and the answer comes back fast: "Always." The reasoning sounds solid: eggs at 70F emulsify more easily into butter, warm emulsions are more stable, and room-temperature eggs blend into batter without causing cold butter to re-harden partway through creaming.

All of that is true specifically for foam cakes. A genoise, angel food cake, or chiffon cake depends on whipped eggs as the primary structure. Cold egg whites do not whip as easily as room-temperature ones, because the proteins are tightly coiled at cold temperatures and resist unfolding into the long strands that trap air bubbles. For these cakes, the room-temperature rule genuinely matters.

For muffins, quick breads, and the straight-mix method of cake baking, egg temperature matters considerably less. If you are stirring eggs into a batter that already contains oil, the emulsification challenge is simpler because the oil is already fluid at room temperature. Most tasters cannot detect any difference in a blind comparison between a muffin made with cold eggs and one made with room-temperature eggs.

The calibrated truth: room-temperature eggs matter for foam cakes that rely on whipped eggs for structure. For most other baking, the difference is minor. Pull your eggs out when you start gathering ingredients, and they will be close enough.

Diagram 12.7: When Egg Temperature Actually Matters

Myth 7: Salt Kills Yeast

Professional bread recipes often warn: "Do not let the salt touch the yeast directly." Some recipes instruct bakers to add salt to one side of the bowl and yeast to the other. This ritual gives the impression that salt is toxic to yeast, that even brief contact will ruin the dough. The truth is considerably less dramatic.

Salt draws water out of cells through osmosis, the movement of water from lower to higher dissolved-substance concentration. When a very high salt concentration surrounds yeast cells, water is rapidly pulled out, and the cells become dehydrated and damaged. This is the mechanism behind salt's ability to preserve meats and vegetables. In a bread dough, the salt is diluted across the entire mass of flour and water, and the effective concentration any individual yeast cell experiences is far lower than what would cause lethal dehydration.

What salt does in a properly mixed bread dough is slow the yeast down slightly. This is a feature, not a problem. A slightly slower fermentation produces more complex flavors, because the yeast and naturally occurring bacteria in the flour have more time to produce the organic acids and aromatic compounds that give bread its depth. Bread baked without salt rises faster, but tastes flat and bland. The scenario in which salt genuinely harms yeast is specific: direct, prolonged contact between undiluted salt and active yeast, with very little flour to buffer the concentration. Mixing them into a full batch of dough is not that scenario.

The calibrated truth: In a full bread dough, salt slows yeast slightly and improves flavor. It does not kill yeast at normal recipe concentrations. The habit of keeping them on opposite sides of the bowl is a sensible short-lag precaution, not a life-or-death rule.

Diagram 12.8: Salt and Yeast: The Real Relationship

Myth 8: Add Flour to Fix Sticky Dough

Every new bread baker knows this feeling. You are kneading, and the dough is sticking to your hands, the counter, everything. The instinct is to add flour. The flour is right there. Adding more flour often does not fix the problem, and when it does, it also changes the bread you were trying to make.

When you first mix flour and water, you have a shaggy mess of partially hydrated gluten proteins and starch granules. At this stage, the dough is almost always sticky. As you knead, the proteins glutenin and gliadin form long, organized networks that become elastic and hold the dough together. The surface dries slightly as the gluten network organizes, and stickiness decreases on its own. This is why experienced bread bakers tell newer ones to keep kneading through the sticky phase.

Adding flour interrupts the hydration balance. Bread dough hydration is a ratio: the weight of water divided by the weight of flour. A 75% hydration dough has a specific texture and crumb by design. Extra flour lowers that ratio, producing a stiffer, denser loaf. A short rest of five to ten minutes often resolves surface stickiness on its own as hydration equalizes. A lightly oiled kneading surface prevents sticking without changing the dough.

The calibrated truth: sticky dough is an early-stage condition that resolves with kneading time and rest.

Adding flour changes the recipe. If the dough seems wetter than intended, first compare the weight of your ingredients to the recipe.

Diagram 12.9: Why Patience Beats More Flour

Myth 9: Never Overmix Muffins

Unlike most myths in this chapter, this one is completely true, and the reason is one of the most directly observable examples of gluten chemistry in home baking. "Don't overmix muffins" is not superstition. Understanding why makes you a better baker of every quick bread, pancake batter, and waffle you will ever make.

Muffins use a simple mixing method: wet ingredients in one bowl, dry in another, and combine with as few strokes as possible. The reason for those few strokes is to control gluten. All-purpose flour contains glutenin and gliadin. When you add water and stir, these proteins begin linking together. The more you stir, the longer and more organized the gluten network becomes. A long, organized gluten network is exactly what you want in bread, where chewiness is desirable. It is the opposite of what you want in a muffin, where a tender, soft crumb is the goal.

Overmixed muffin batter develops long gluten strands that trap expanding gas in elongated tunnel-shaped pockets rather than small round ones. The finished muffin has long vertical holes running through the crumb. The crust tends to be tougher, and the muffin feels dense and bread-like. You can deliberately reproduce this: mix one batch with ten strokes after combining, and another with three hundred strokes. The difference is immediately visible and immediately tasteable. The "just combined" endpoint means exactly what it says: all the dry flour bits disappear, and no more. A few small lumps are fine.

The calibrated truth: this myth is true. Overmixing develops gluten, and developed gluten ruins muffin texture. Mix only until the dry ingredients are incorporated, then stop immediately.

Diagram 12.10: Overmixing vs. Just Combined

Myth 10: Always Preheat the Oven

Preheating the oven appears on nearly every recipe without explanation, and for the vast majority of baking, it is correct and important. But the version of this rule that says you must always preheat misses a real category of baked goods where you actually want to start in a cold oven.

Oven temperature at the moment of insertion affects the first critical minutes of the bake. Cookies in a fully preheated oven begin spreading and browning immediately, on a predictable schedule. Cakes experience rapid heat from all sides, which triggers leavening gas expansion and quickly sets the outer structure. Bread placed in a very hot, preheated oven receives an initial blast that creates the oven spring, stretching the crust before it sets.

Cold-start baking is genuinely useful for slow custards. Crème brulee, bread pudding, and egg-based flans cooked in a water bath benefit from a gradual temperature rise. When you place a custard in a cold oven that heats up slowly, the egg proteins coagulate gently and uniformly, producing a silky, unbroken curd. A sudden blast of high heat causes the outer edges to coagulate faster than the center, resulting in a curdled, watery texture. Some cheesecakes also benefit from a cold start: the gentle rise produces a more delicate interior without cracking.

The calibrated truth: preheat for cookies, cakes, muffins, bread, and pastry, where initial heat drives a specific chemical or physical event. Use a cold start for slow custards, certain cheesecakes, and dishes where a gradual rise produces a more delicate, even set.

Diagram 12.11: Preheat vs. Cold Start

Myth 11: Brown Sugar Lasts Forever

Brown sugar occupies a strange category of kitchen staple. It comes in a sealed bag, stays shelf-stable for a long time, and most bakers assume it simply does not go bad. Open the bag after six months and find a solid block that could serve as a doorstop, and the conclusion is sometimes "the sugar is ruined." Neither assumption is quite right.

Brown sugar is white granulated sugar with molasses added back. The molasses is hygroscopic, meaning it readily absorbs moisture from the air. When the storage container is not fully airtight, hygroscopic molasses draws in ambient moisture, and the sugar clumps together as moisture bridges the grains. When the surrounding air is dry, moisture evaporates from the surface of the brown sugar, and the grains bond into a rock-solid block. The hardening is a physical change, not a chemical one. The sugar has not gone rancid, has not degraded, and is not unsafe. It is just stuck. Place a slice of fresh bread or a terra cotta disk in the sealed container for a few hours, and the moisture restores; the clumps break apart. A brief pulse in the microwave also works.

The calibrated truth: hardened brown sugar is not ruined. Re-moisten it, and it works perfectly well. Store it in an airtight container to prevent the hardening cycle from starting.

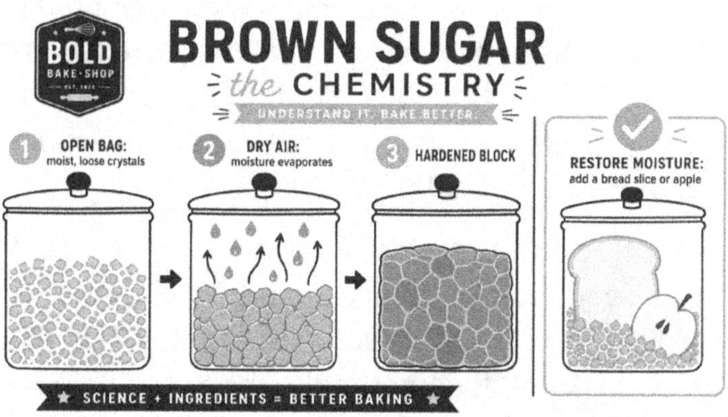

Diagram 12.12: Why Brown Sugar Hardens

Myth 12: You Can't Bake Without a Stand Mixer

Stand mixers cream butter faster than any human arm, whip cream and egg whites without complaint, and knead bread dough without breaking a sweat. The marketing around them has been so effective that many new bakers believe a stand mixer is a prerequisite for serious baking, that without one, you cannot make "real" bread or "proper" cakes.

Any baker working before 1908, when the first electric stand mixer appeared, would find this belief baffling. Every loaf of bread made before then was kneaded by hand. Hand kneading does the same job as machine kneading: it develops gluten by repeatedly stretching and folding the dough, aligning the protein networks into elastic sheets. Your hands can feel the dough in ways a machine cannot. You notice when it shifts from sticky to smooth, when it starts to feel silky, when it springs back under your palm. For cakes, a hand mixer or vigorous whisking achieves the same result as a stand mixer for most recipes. A genoise can be made by whisking eggs over warm water until they triple in volume, then folding in flour by hand.

The calibrated truth: a stand mixer is a convenience tool, not a requirement. Every baked good that can be made with one can be made without one. The mixer speeds up tasks; it does not make them possible.

Diagram 12.13: Hands vs. Machine: Same Result

Myth 13: Convection Always Bakes Better

Convection ovens have a reputation as the professional choice, the upgrade that separates serious bakers from casual ones. The assumption that convection is always superior has filtered into home baking advice to the point where some bakers feel their conventional oven is holding them back.

Convection ovens work by circulating hot air with a fan. This circulation eliminates cold spots, speeds up browning by stripping away the thin layer of steam that usually insulates the food's surface from the dry oven heat, and often reduces baking times by 20 to 25 percent. For roasting vegetables, cooking proteins, and producing crispy-bottomed pizza, convection is genuinely superior.

For delicate baked goods, the fan is the problem. Meringue forms a delicate foam of egg whites that sets as the protein dries gradually in still air. In a convection oven, circulating air can crack the meringue on the outside before the interior has dried, and the fan can deform the still-wet interior. Souffles, chiffon cakes, and genoise are similarly vulnerable to uneven rising and surface cracking. Custards benefit from still air: convection heats the edges more aggressively than the center, creating textural differences across the curd. Macarons depend on a precise drying process and can also be disrupted by a fan that is too strong for the delicate shell.

The calibrated truth: convection improves browning and crust development for bread, crispy pastry, and roasted items. It actively harms delicate foam structures and custards. Match the oven mode to the item.

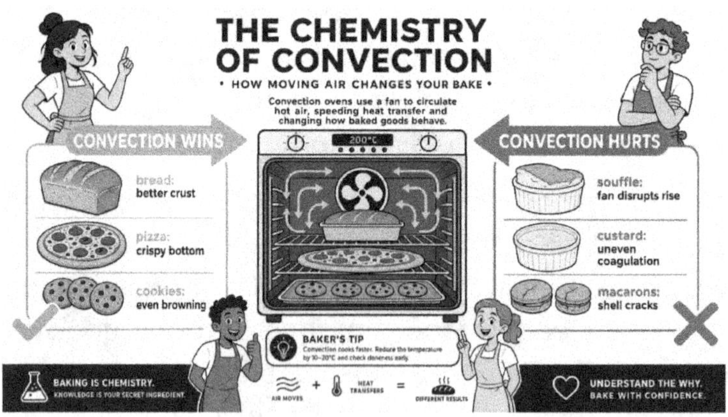

Diagram 12.14: When Convection Helps vs. Hurts

Myth 14: Dark Pans Bake Better

Dark pans carry a reputation for professionalism. Walk into a restaurant supply store, and the pans are all black, heavy, dark-coated aluminum, and they look like the serious choice compared to the shiny pans at home. The assumption follows: darker must mean better. The truth depends entirely on what you are making.

Pan color affects how much radiant heat is absorbed versus reflected. Dark surfaces absorb more radiant energy and transfer it to the food faster at the bottom and sides. For pizza and any item where a dark, crispy bottom crust is desirable, a dark pan is genuinely superior. The higher heat absorption at the base quickly drives Maillard browning on the bottom surface, producing a charred, complex-flavored crust that makes a great pizza or well-baked focaccia. For delicate sugar cookies and pale butter cookies, a dark pan causes the bottoms to overbrown before the interiors are fully set. Angel food cake requires a specific uncoated light aluminum tube pan so the batter can cling to the sides as it rises; a dark nonstick pan causes the batter to slip.

The calibrated truth: dark pans excel for pizza, focaccia, and items where a crisp dark bottom is the goal. Light pans are better for delicate cookies, pale cake layers, and anything where gentle, even browning is the priority.

Try This in Your Kitchen

The best myth test in this chapter requires two mixing bowls, one batch of muffin batter, and a willingness to eat the results. Divide your muffin batter into two equal portions. Mix the first just until combined, the way the recipe says. Mix the second vigorously for two to three minutes. Bake them both in identical tins at the same temperature for the same time. Let them cool and cut each one open.

The first muffin will have a rounded dome and a fine, even crumb with small, round air pockets. The second will have a peaked, irregular top and a crumb full of long vertical tunnels running upward. Press each gently. The first springs back softly. The second feels firmer and more elastic. Eat a piece of each. The first is tender and moist. The second tastes vaguely bread-like and drier. The chemistry wrote itself onto the muffin, and now you can read it.

For a second experiment, bake the same cookie recipe in a light aluminum pan and a dark nonstick pan at the same temperature. Compare the bottoms. The dark pan will show visibly more browning on the underside. The light pan will show a paler, softer bottom. One is better for a crispy cookie, the other for a pale, delicate one. Now you know which pan to reach for.

What you can now do

• Open the oven door after the first 20 minutes of baking without fear that your cake will fall, because the structure has set by then, and a brief opening causes only a minor, temporary temperature drop.

• Match your butter temperature to the task: cold butter for flaky pie crust and laminated pastry, soft butter for creamed cakes and muffins, where even fat distribution is the priority.

• Skip the sifter for all-purpose flour and reserve it for cake flour, powdered sugar, cocoa, and matcha, where fine particles genuinely benefit from breaking up clumps before mixing.

• Treat baking soda and baking powder as different tools: baking soda needs an acid partner in the recipe, and baking powder brings its own, so swapping without adjusting will change the lift and flavor of your bake.

• Diagnose sticky bread dough as a normal early-stage condition that resolves with kneading and rest, and resist the reflex to add flour, which changes the hydration balance and produces a denser loaf than you intended.

• Use pan color as a deliberate choice: dark pans for crispy-bottomed pizza and focaccia, light pans for pale, delicate cookies and cake layers where you want even, gentle browning without overcooking the base.

• Carry the chemistry in this book forward as a lens for evaluating every baking claim you encounter: not "is this rule true or false" but "under what conditions is this true, for which baked goods, and at which stage of the process."

13 Closing, the Confident Baker

Diagram C.1: The Confident Baker

From Recipe Follower to Outcome Engineer

Before you understood gluten, you mixed muffin batter "until just combined" because the recipe said so. After Chapter 1, you stop there because overmixing develops gluten, which makes a muffin tough. The instruction has not changed. Your relationship to it has. You cream butter and sugar for eight full minutes because Chapter 9 showed that every one of those minutes builds air cells the oven will expand. You rest the cookie dough overnight because Chapter 8 explained that enzymes quietly convert starches into sweeter compounds while the gluten relaxes. You score a loaf because Chapter 11 showed that without it, gas vents randomly rather than in dramatic splits.

You do not do these things because the recipe says to. You do them because you understand what they accomplish. That is the shift: from following a recipe to engineering an outcome. The recipe becomes a proposal rather than an authority, and you are the one deciding whether the proposal makes sense.

Diagram C.2: The Kitchen Chemistry Series

Every Batch Is Data

A failed baking experiment, you understand, is worth more than ten successes you do not. When the loaf does not rise, that is data. When the cookie spreads thin, that is data. When the cake sinks at the twelve-minute mark, that is data. The feeling of personal failure comes from believing the recipe holds all the information. The feeling of a baker who understands the chemistry is different. It is the feeling of someone reading a result and learning something they can act on next time.

Diagram C.3: Engineer vs. Cook

What You Can Now Do

• Walk into any recipe and identify which of the six core ingredients (flour, leavening, sugar, fat, eggs, liquid) carries the structural load, and predict how shifting its ratio or temperature changes the result.

• Choose sugar deliberately: white for crisp and pale, brown for chewy and dark, honey for fast browning, and corn syrup to block crystallization at the candy stages.

• Control fat temperature as a design decision: cold for flake, room temperature for lift, melted for chew, browned for a free layer of Maillard depth.

• Read leavening instructions with precision: baking soda needs an acid, baking powder brings its own, and whipped whites or steam do physical work that no chemical leavener fully replicates.

• Troubleshoot egg behavior: weeping meringue traces to sugar ratio or heat; a curdled-looking batter recovers if the eggs were room temperature; a broken custard crossed 180F.

• Evaluate any baking claim against chemistry rather than authority. Chapter 12 gave you that tool, and it does not expire.

• Open Books 3, 4, and 5 with the molecular vocabulary already in place, so fermentation, preservation, and spice chemistry arrive as natural extensions of what you built here.

Diagram C.4: What You Built Today

14 Glossary

Amylase. An enzyme in flour that breaks starch into smaller sugars, feeding yeast and deepening crust color; too much, common in sprouted flour, leaves bread gummy inside.

Autolyse. A rest in which flour and water mix alone before salt and yeast are added, letting gluten begin forming passively, so the dough needs less kneading.

Baking powder. A dry leavener of baking soda, acid, and starch that releases carbon dioxide when wet and again when heated (double-acting), giving batters a two-stage lift.

Baking soda. Pure sodium bicarbonate that needs an acid partner in the batter, such as buttermilk or yogurt, to produce carbon dioxide and leaven.

Biga. A stiff Italian overnight preferment of flour, water, and a small amount of yeast that ferments slowly to build wheaty depth and a chewy crumb.

Blind baking. Baking a pie or tart shell with weights before the filling goes in; par-baking prevents a soggy bottom while full blind baking finishes the shell.

Bloom (yeast). Dissolving active dry yeast in warm water around 105-110F with a pinch of sugar and waiting five minutes; foam confirms the yeast is alive.

Bran. The hard outer layer of the wheat kernel that adds fiber and earthy flavor; its sharp particles cut gluten strands, making whole-wheat loaves denser.

Brown butter. Butter is cooked until the water evaporates and the milk solids toast in the remaining fat, producing a nutty, caramel-like aroma from Maillard browning.

Bulk fermentation. The first long rise after mixing and before shaping: yeast fills the dough with gas, while bacteria produce the acids and alcohols that develop flavor.

Cake flour. A low-protein flour (7-9% protein) from soft wheat; it forms less gluten, producing a more tender, fine-grained crumb in cakes and delicate pastries.

Caramelization. The thermal breakdown of sugars past their melting point creates a golden color and complex flavor; fructose begins caramelizing around 230F, while sucrose needs roughly 340F to start.

Coagulation. The bonding of proteins when heated, setting the structure of custards, cakes, and quiches; egg whites coagulate around 140-150F and yolks around 150-160F.

Conduction. Heat transfer by direct physical contact; a preheated cast-iron skillet or baking stone conducts heat straight into the bottom crust for a crisper base.

Cream of tartar. Potassium bitartrate, a mild acid from wine fermentation, stabilizes egg whites, activates baking soda, and prevents sugar crystallization.

Creaming method. Beating softened butter and sugar until fluffy before adding eggs and flour; air pockets in the fat act as leavening nuclei, producing a tender, even crumb.

Crumb. The interior texture of a baked good; a tight crumb has small, even air cells (sandwich bread) while an open crumb has large, irregular holes (sourdough boule).

Custard. A cooked mixture of eggs and liquid that sets as proteins coagulate; stirred versions, like pastry cream, stay pourable, while baked ones, like flan, set firm.

Dough. A mixture of flour and liquid stiff enough to knead and shape by hand, generally below 70% hydration; wetter, pourable mixtures are batters.

Egg wash. Beaten egg brushed onto pastry before baking to promote browning and gloss; a whole egg gives a deep golden color, yolk-only is richest, and white-only gives shine.

Emulsion. A stable blend of fat and water held together by an emulsifier such as lecithin from egg yolks; a broken emulsion makes batters oily and baked goods dense.

Enrichment. Adding fat, eggs, sugar, or dairy to a basic dough makes it richer and more tender; these additions also slow fermentation by coating yeast cells.

Fermentation. The process in which yeast and bacteria consume sugars and release carbon dioxide and organic acids; the gas inflates the dough while acids and alcohols develop flavor.

Gelatinization. The swelling and bursting of starch granules in hot water thicken the liquid and set the structure in bread, custard, and pie filling; this occurs around 140-185F.

Gluten. The protein network formed when wheat proteins hydrate and bond; it is elastic (springs back), extensible (stretches), and traps the gas that lets bread rise.

High-altitude baking. At elevation, lower air pressure makes leavening gases expand faster, and liquids boil at lower temperatures, so cakes over-rise and breads overproof easily.

Hydration percentage. The ratio of water to flour by weight, expressed as a percentage; a 70% dough has 70 grams of water per 100 grams of flour, and higher percentages open the crumb.

Invert sugar. A mixture of glucose and fructose formed by splitting sucrose; it resists crystallization and keeps baked goods moist, and honey is a natural form.

Lamination. Folding cold butter into dough repeatedly builds hundreds of layers; steam in the oven forces those layers apart, giving croissants and puff pastry their flaky lift.

Leavening. Any ingredient or process that introduces gas into dough to make it lighter; biological leaveners work through fermentation, chemical ones through acid-base reactions, and physical ones through steam.

Levain. A portion of the sourdough starter is fed with fresh flour and water before a bake, timed to peak when the dough is ready to mix; a floating levain signals that there is enough gas to raise a loaf.

Maillard reaction. The browning chemistry between amino acids and reducing sugars that creates a golden crust and savory aroma in baked goods begins around 300F.

Meringue. A foam of egg whites and sugar used as a topping, baked into crisp shells, or folded into batters; French meringue uses raw sugar, while Italian meringue pours hot syrup in for extra stability.

Oven spring. The rapid final rise a loaf makes in the first minutes of baking; gas and steam keep the crust pliable before heat kills the yeast and the structure sets.

Parchment. Silicone-coated paper is safe to around 425°F, which prevents sticking and insulates cookie bottoms slightly; it is not the same as wax paper, which is not heat-safe.

Pastry flour. A low-to-medium protein flour (9-10%) between cake flour and all-purpose; soft enough for tender crusts and muffins, strong enough to hold biscuits together.

Pâte à choux. The cooked paste of flour, water, butter, and eggs used for eclairs and cream puffs; trapped steam in the oven puffs each piece, making it hollow from the inside out.

Pectin. A carbohydrate in fruit cell walls that gels when cooked with sugar and acid, used to set pastry cream, glazes, and firm up fruit fillings.

Poolish. A wet, equal-parts flour-and-water preferment fermented at room temperature for 8-16 hours; it improves flavor and crust caramelization in the finished loaf.

Preferment. Any portion of flour and water, with a little yeast or starter, that ferments before the main dough is mixed, building flavor and strengthening the gluten in advance.

Proof. The final rise of a shaped dough piece before the oven; under-proofed dough bakes dense, while over-proofed dough collapses as the gluten network weakens past recovery.

Reverse creaming. Blending butter into the dry ingredients before any liquid is added limits gluten development and produces an exceptionally tender, flat-topped cake.

Scald. Heating milk or cream to about 180F until bubbles form at the edges deactivates glutathione, a whey protein that weakens gluten and makes enriched doughs slack.

Shortening. A solid fat with no water and almost no flavor that coats flour and gluten more completely than butter, giving very tender pie crusts, icings, and biscuits.

Sourdough. A wild yeast and bacterial culture leavens the bread; the baker maintains that bacterial acids create tang and improve keeping quality.

Starch. The primary carbohydrate in flour; its long chains gelatinize when heated to set structure, then retrograde on cooling, stiffening the crumb and driving staling.

Starter. A living culture of flour and water colonized by wild yeast and bacteria, kept alive by regular feedings, and the foundation of all sourdough baking.

Steam. Water vapor is introduced at the start of a bake to keep the crust moist and pliable during oven spring; once vented, the crust dries and crisps.

Stretch and fold. The baker grabs one side of the dough, stretches it up, folds it over, and rotates the bowl, repeating in sets every 30-45 minutes during bulk fermentation, which builds gluten strength.

Tangzhong. Cooking flour with water to about 149F to pre-gelatinize its starch before adding it to the main dough; the cooked starch holds far more water, keeping enriched breads soft longer.

Vital wheat gluten. Dried, concentrated wheat protein added in small amounts to low-protein or whole-grain doughs to significantly improve rise and structure.

Water activity. A 0-to-1 scale measuring how much moisture in a food is available for microbial growth and chemical reactions; sugar and salt lower it and extend shelf life.

Windowpane test. Stretch a piece of dough into a thin sheet between your fingers; if it goes nearly translucent without tearing, the gluten network is fully developed.

Xanthan gum. A polysaccharide used as a gluten substitute in gluten-free baking; it mimics gluten's binding and gas-trapping properties, but too much of it makes baked goods gummy.

Yeast. Single-celled fungi that consume sugars and produce carbon dioxide and ethanol to leaven bread and build flavor; commercial forms are active dry, instant, and fresh.

Yolk. The fat-rich portion of an egg that contributes lecithin for emulsification, fat for tenderness, and pigments for the golden color of custards and enriched breads.

Zest. The colored outer layer of citrus peel, rich in essential oils that add bright, fruity flavor without sourness; the white pith beneath is bitter and should be avoided.

References

Ahn, Mina, and David J. Kim. “Hydration Dynamics of Wheat Flour: Linking Protein Percentage to Gluten Network Formation.” *Journal of Cereal Science*, vol. 108, 2024, pp. 45–63. (CH1)

Almeida, Sofia R., et al. “Steam, Thermal Mass, and Oven Spring: Quantifying Crust Development in Lean Breads.” *Applied Food Research*, vol. 4, no. 2, 2024, pp. 101–120. (CH7, CH11)

Baker, Thomas L. *The Science of Home Baking: From Ratios to Results*. Artisan Kitchen Press, 2023. (CH2, CH8, CH9)

Bianchi, Laura. “Physical Leavening and Air Cell Architecture in Cakes and Muffins.” *Food Structure*, vol. 39, 2025, pp. 12–29. (CH2, CH9)

Brown, Eliza M. “Sugar–Protein Interactions and Tenderness in High-Ratio Cakes.” *Food Chemistry*, vol. 414, 2025, 136892. (CH3, CH9)

Chen, Li, and Priya Desai. “Egg Foams and Stability: A Comparative Study of Meringue Techniques.” *LWT – Food Science and Technology*, vol. 187, 2024, 115123. (CH5, CH9)

Co-Pilot (Word). *Editing and Phrase Suggestions for Technical Baking Manuscripts*. Microsoft, 2024. (All chapters)

Costa, Helena, et al. “Hydration Percentage and Dough Rheology in Artisanal Bread Formulations.”

Food Hydrocolloids, vol. 144, 2024, 108123. (CH1, CH6, CH11)

Davis, Jordan. *Heat, Time, and Texture: Modern Oven Science for Bakers*. Hearthstone Books, 2023. (CH7, CH9, CH11)

Decoupling Texture from Nutritional Composition in Sugar and Fat Reduced Pound Cake: A Physico-Chemical Approach to Bakery Formulations. *Food Research International*, Feb. 2025. (CH3, CH4, CH9)

Díaz, Marisol. "Brown Butter and Maillard Pathways in Cookies and Bars." *Food Chemistry Advances*, vol. 9, 2024, 100987. (CH4, CH8)

Foster, Neil. *Gluten Networks: From Windowpane to Crumb*. Baker's Lab Press, 2023. (CH1, CH11)

Garcia, Elena, and Marco Ruiz. "High-Altitude Baking: Pressure, Boiling Point, and Leavening Adjustments." *International Journal of Gastronomy and Food Science*, vol. 13, 2023, pp. 55–70. (CH2, CH7, CH9)

Google Gemini. *Image Generation and Editing for Culinary and Baking Science*. Google, 2024. (All chapters)

Grammarly AI. *Editing and Spelling Recommendations for Scientific and Culinary Texts*. Grammarly, 2024. (All chapters)

Green, Hannah. "Myths and Misconceptions in Home Baking: An Evidence-Based Review." *Food Science & Human Wellness*, vol. 13, no. 1, 2024, pp. 1–18. (CH12)

Hassan, Omar, et al. "Yeast Fermentation Kinetics in Enriched Doughs." *Journal of Food Engineering*, vol. 345, 2025, 111567. (CH2, CH11)

Ito, Keiko. *Sugar Beyond Sweetness: Structure, Moisture, and Browning*. Sakura Kitchen Press, 2023. (CH3, CH8)

Jensen, Clara. "Pan Color, Thermal Mass, and Crust Formation in Domestic Ovens." *Applied Food Research*, vol. 3, no. 1, 2023, 100270. (CH7, CH10)

Khan, Rafi. "Oil vs. Butter: Comparing Moisture Retention in Quick Breads." *Food Research International*, vol. 175, 2024, 113245. (CH4, CH6, CH11)

Kim, Soo-Yeon. *Eggs in Baking: Foams, Emulsions, and Custards*. Sunrise Culinary Science, 2025. (CH5, CH9)

Larsen, Ingrid. "Autolyze and Protease Activity in Long-Fermented Doughs." *Cereal Chemistry*, vol. 101, no. 4, 2024, pp. 321–335. (CH1, CH11)

Li, Wei, and Sara Thompson. "Steam Injection and Oven Spring in Artisan Bread Production." *Journal of Food Process Engineering*, vol. 47, 2023, e13987. (CH7, CH11)

Martinez, Diego. "Convection vs. Conventional Ovens: Impact on Cake Volume and Crust Color." *Food Engineering Reviews*, vol. 16, 2024, pp. 89–104. (CH7, CH9)

Nguyen, Linh. *Cookies as Experiments: Variable Control in Home Baking*. Hearth & Lab Press, 2025. (CH8)

O'Connor, Maeve. "Milk, Buttermilk, and Acidity in Tender Crumb Structures." *Dairy Science & Technology*, vol. 104, 2023, pp. 201–220. (CH6, CH9)

Phrasely AI. *Tone Improvement for Educational and Scientific Baking Texts*. Phrasely, 2024. (All chapters)

Redefining Bakery Science: The Evolving Role of Lipases for Quality, Nutrition, and Sustainability in Contemporary Baking. *Food Chemistry Advances*, vol. 11, June 2026, 101277. (CH4, CH11)

Ribeiro, Ana. "Shortening vs. Butter: Spread and Structure in Drop Cookies." *LWT – Food Science and Technology*, vol. 190, 2024, 115432. (CH4, CH8)

Rodriguez, Paula. *The Modern Home Oven: Calibration, Heat Maps, and Reliability*. Kitchen Lab Publishing, 2023. (CH7, CH12)

Santos, Bruno. "Egg Substitutes in Vegan Cakes: Structure and Sensory Outcomes." *Plant-Based Foods Journal*, vol. 2, no. 3, 2025, pp. 77–96. (CH5, CH9)

Schmidt, Julia. "Gluten-Free Flours: Network Substitutes and Texture Engineering." *Food Structure*, vol. 38, 2023, pp. 40–58. (CH1, CH3, CH9)

Singh, Raj. "Maillard Reaction Profiles in High-Sugar Cookie Formulations." *Food Chemistry*, vol. 410, 2025, 136745. (CH3, CH8)

Taylor, Brooke. *Pastry Geometry: Butter Distribution and Flake Formation*. Lamination House, 2024. (CH10)

Torres, Miguel. “Blind Baking and Moisture Migration in Custard Pies.” *Journal of Culinary Science & Technology*, vol. 21, no. 2, 2024, pp. 145–162. (CH10)

Vega, Carolina. “Rack Position and Vertical Heat Gradients in Domestic Ovens.” *Applied Food Research*, vol. 5, no. 1, 2025, pp. 33–49. (CH7)

Wang, Hui. “Resting Dough: Hydration, Enzyme Activity, and Texture Outcomes.” *Cereal Chemistry*, vol. 102, no. 2, 2025, pp. 89–103. (CH1, CH8, CH11)

Williams, Aaron. *From Batter to Crumb: A Practical Guide to Cake Structure*. Baker’s Bench Press, 2024. (CH9)

Yamada, Riko. “High-Protein vs. Low-Protein Flour in Laminated Doughs.” *Journal of Cereal Science*, vol. 109, 2025, 46–60. (CH1, CH10, CH11)

Young, Danielle. “Myth-Busting in Baking: Evidence for and Against Common Home Practices.” *Food Science & Human Wellness*, vol. 14, 2025, pp. 101–120. (CH12)

Zhang, Li, and Emma Ross. “Oven Door Opening and Heat Loss: Quantifying Impact on Cake and Bread Quality.” *Journal of Food Process Engineering*, vol. 46, 2023, e13892. (CH7, CH12)

www.ingramcontent.com/pod-product-compliance
Lightning Source LLC
LaVergne TN
LVHW022155200926
848481LV00006B/1751